SCIENCE ADVENTURERS

SPACE SCIENTISTS

BY KATHRYN HULICK

CONTENT CONSULTANT

Daniel C. Jacobs
Assistant Professor
School of Earth and Space Exploration
Arizona State University

Essential Library
An Imprint of Abdo Publishing
abdobooks.com

ABDOBOOKS.COM

Published by Abdo Publishing, a division of ABDO, PO Box 398166, Minneapolis, Minnesota 55439. Copyright © 2020 by Abdo Consulting Group, Inc. International copyrights reserved in all countries. No part of this book may be reproduced in any form without written permission from the publisher. Essential Library™ is a trademark and logo of Abdo Publishing.

Printed in the United States of America, North Mankato, Minnesota.
092019
012020

THIS BOOK CONTAINS RECYCLED MATERIALS

Cover Photos: NASA (front); Shutterstock Images (back)
Interior Photos: JPL-Caltech/NASA, 4–5, 6–7, 51, 52–53, 61, 64–65, 70–71, 78–79; JPL-Caltech/MSSS/NASA, 10–11; NASA, 14, 23; New York Public Library/Science Source, 16–17; MSFC/NASA, 19, 24, 30; Bill Ingalls/NASA, 26–27, 73; JSC/NASA, 29, 34, 36, 39; JPL/NASA, 40–41, 90–91; JPL/Space Science Institute/NASA, 44; KSC/NASA, 46, 68; Johns Hopkins University Applied Physics Laboratory/Southwest Research Institute/NASA, 49; ESA/ATG Medialab/Science Source, 55; Rosetta/Philae/CIVA/ESA, 56–57; Kyodo/AP Images, 62; European Southern Observatory/Science Source, 67; ESO, 74; Phil Degginger/Science Source, 77; NASA Goddard, 82; Caltech/MIT/LIGO Lab/Science Source, 86–87; Mark Garlick/Science Source, 89; SpaceX, 93; Refugio Ruiz/AP Images, 94; Gary Gershoff/WireImage/Getty Images, 98

Editor: Arnold Ringstad
Series Designer: Laura Graphenteen

LIBRARY OF CONGRESS CONTROL NUMBER: 2019942079

PUBLISHER'S CATALOGING-IN-PUBLICATION DATA

Names: Hulick, Kathryn, author.
Title: Space scientists / by Kathryn Hulick
Description: Minneapolis, Minnesota : Abdo Publishing, 2020 | Series: Science adventurers | Includes online resources and index.
Identifiers: ISBN 9781532190353 (lib. bdg.) | ISBN 9781532176203 (ebook)
Subjects: LCSH: Space sciences--Juvenile literature. | Aerospace engineering--Juvenile literature. | Scientists--Juvenile literature. | Discovery and exploration--Juvenile literature. | Adventure and adventurers--Juvenile literature.
Classification: DDC 629.409--dc23

CONTENTS

CHAPTER ONE
EXPLORING MARS 4

CHAPTER TWO
THE RACE TO SPACE 16

CHAPTER THREE
LIFE AS AN ASTRONAUT 26

CHAPTER FOUR
MYSTERIOUS MOONS 40

CHAPTER FIVE
CATCHING COMETS AND ASTEROIDS 52

CHAPTER SIX
SEARCHING FOR OTHER EARTHS 64

CHAPTER SEVEN
THE EXPANDING UNIVERSE 78

CHAPTER EIGHT
INTO THE UNKNOWN 90

ESSENTIAL FACTS 100
GLOSSARY 102
ADDITIONAL RESOURCES 104
SOURCE NOTES 106
INDEX 110
ABOUT THE AUTHOR 112

CHAPTER ONE

EXPLORING MARS

A seasoned explorer scoots across the slope of a mountain on an alien world, scanning the desolate landscape with cameras and sensors. This explorer isn't human. It's *Curiosity*, a robotic rover the size of a car. For more than nine years, *Curiosity* has poked, prodded, photographed, zapped, and tested as it has rolled its way across Mars. Meanwhile, people on Earth working with *Curiosity* have been analyzing the information the robot collects. They're investigating the question of whether Mars has or ever had the right conditions to support life. But before that could happen, the robot explorer had to land safely on another planet. That was no easy task.

TOUCHDOWN

On the night of August 5, 2012, after a 253-day journey through space, the spacecraft carrying *Curiosity* zoomed into the Martian atmosphere at a speed

The *Curiosity* rover, a robotic vehicle the size of a small car, has been exploring Mars since 2012.

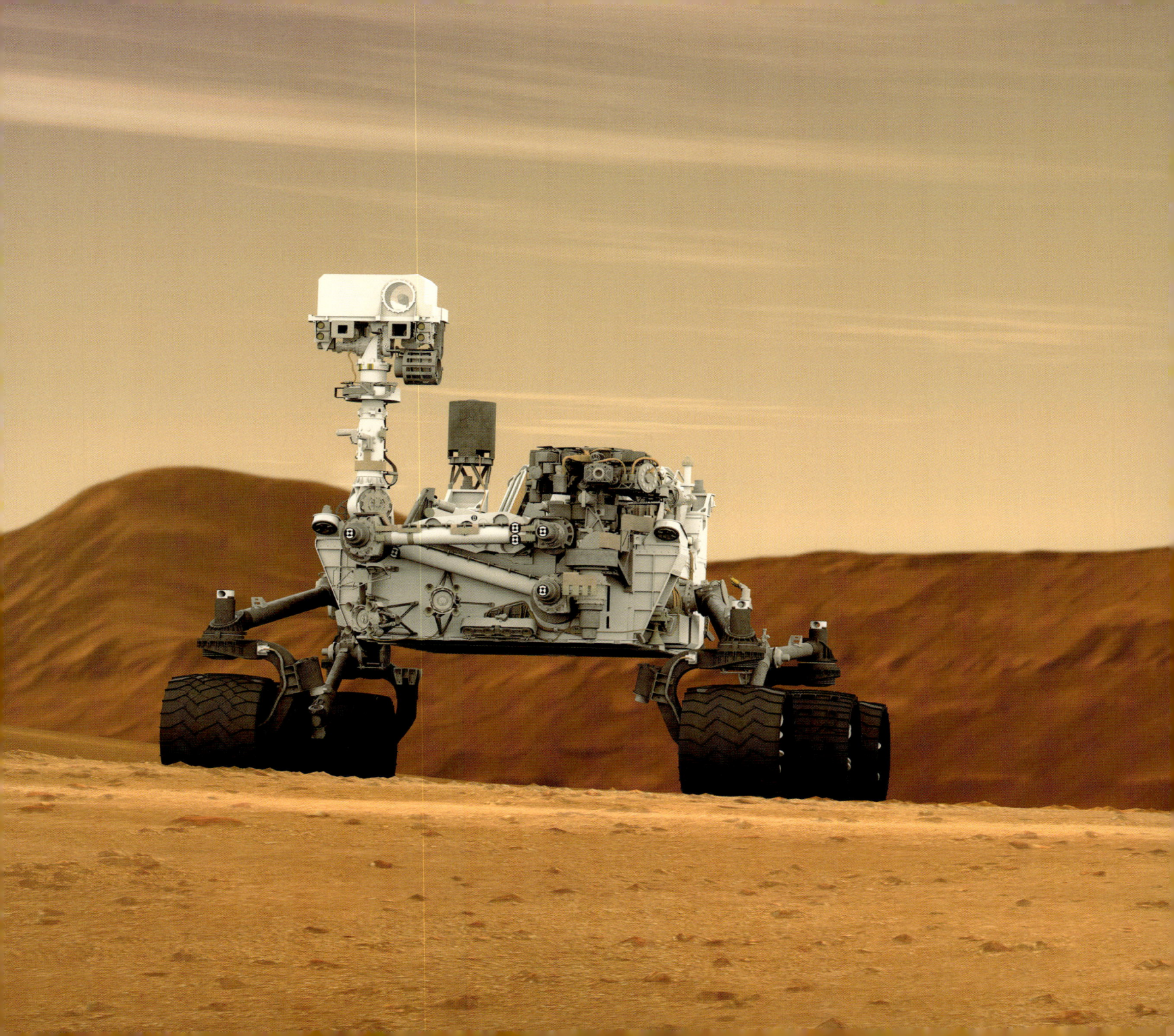

of more than 13,000 miles per hour (21,000 km/h).[1] The entire landing process would take about seven minutes. But it took about 14 minutes for signals to travel between Mars and Earth. That meant the computers on the spacecraft had to accomplish the landing on their own, without any real-time help from the team on Earth. National Aeronautics and Space Administration (NASA) engineers dubbed the landing sequence "seven minutes of terror."[2]

First, the spacecraft had to withstand the extreme heat generated when it began hitting the atmosphere. A heat shield would help with that. Then a parachute had to open, slowing the descent. Next the spacecraft had to ditch the heat shield and parachute and start firing eight rocket engines toward the ground. This would slow down the craft even further. Finally, a device called a sky crane was supposed to lower *Curiosity* down on cables,

Curiosity's daring sky crane landing system worked perfectly, bringing the rover to a safe touchdown.

gently place the rover on the ground, release itself, and fly away for a crash landing at a safe distance.

While all this was going on, a room full of NASA scientists and engineers nervously waited for the messages that would tell them whether the arrival had gone as planned. Some had been working on the project for seven years or more. As they waited, they paced or wrung their hands. Some felt sick to their stomachs. Adam Steltzner, the engineer in charge of the landing, hadn't slept well in weeks. Finally, the announcement came: "Touchdown confirmed. We are safe on Mars."[3] The room erupted in cheers and applause. "I felt strangely numb, exhilarated, and slightly in disbelief," Steltzner said afterward. "You work so many years of your life, so hard, on something that seems crazy even to us at times, it looks risky in a very visceral way, and then to have it just work out. . . . To be honest I'm still a little in shock that it worked."[4]

Curiosity wasn't the first robot to land on Mars, but it was the biggest. It also carried more advanced equipment than earlier robots. It soon began to explore exciting regions of the planet, Gale Crater and Mount Sharp.

SPIRIT AND OPPORTUNITY

In 2004, two robotic rovers, *Spirit* and *Opportunity*, landed on opposite sides of Mars. They both collected evidence showing that Mars was a wetter and warmer world in the distant past. Their main missions were only supposed to last 90 days, but both robots rolled on for much longer. *Spirit* stopped working after it got stuck in soft sand in 2010. And in 2018, a sandstorm crippled *Opportunity*. The robot could no longer charge its batteries, and it stopped communicating with Earth. It had lasted more than 5,000 days—more than 50 times its planned life span.[5]

WATERY CLUES

Scientists chose these areas for a good reason. Satellite images had revealed that liquid water may have once flowed through channels there. And water is one of the key ingredients that life as we know it needs to survive. A few weeks after *Curiosity* landed, the rover found streambeds and a lakebed. Both had carried flowing, liquid water billions of years ago. Back then, the planet had a much warmer climate. In the ancient mud at the bottom of the dried-up lake, *Curiosity* also found some of the chemical ingredients that living cells require to survive, including sulfur, nitrogen, hydrogen, oxygen, phosphorus, and carbon.

Earth is the only place in the universe where life is known to exist. But long ago, this Martian lake could have supported life. And it likely stayed that way for a billion years or longer. That means it could have been home to very small, very simple life forms called microbes. "It is exciting to think that billions of years ago, ancient microbial life may have existed in the lake's calm waters, converting a rich array of elements into energy," said planetary scientist Sanjeev Gupta of Imperial College London.[6]

The *Curiosity* team posts discoveries, pictures of Mars, and even selfies from the rover on Twitter.

LIVING ON MARS TIME

Mars scientists may not live on Mars. But some live on Mars time. One day on Earth lasts 24 hours. This is the amount of time it takes for the planet to rotate once on its axis. On Mars, one complete rotation takes 24 hours and 40 minutes, a time period called a sol. When the Mars rovers *Spirit* and *Opportunity* landed in 2004, an entire team of scientists aligned their schedules to the Martian sol. That meant their bedtimes and meal times had to shift 40 minutes later every single day.[7]

Today, most of the water on Mars is trapped inside ice or floating in the air as vapor. But liquid water may still flow in some places. Several satellites that orbit the planet have photographed dark streaks that show up regularly on some slopes on Mars. They may be seeps of liquid water. Or they could just be sand. In 2018, the *Mars Express* spacecraft used radar to peer beneath a frozen sheet of ice at the planet's south pole. It found a liquid lake buried there. The lake is likely frigid and very salty. But if there's one underground lake, there could be more. "It is an exciting prospect to think there could be more of these underground pockets of water elsewhere, yet to be discovered," said Roberto Orosei of the National Institute for Astrophysics in Italy, who led the *Mars Express* investigation.[8]

Curiosity's sharp images from the Martian surface help scientists learn more about the planet.

Curiosity uses a laser to vaporize and study rocks.

MARS MISHAPS

Curiosity landed safely. But several other Mars missions ended in failure. The NASA *Mars Polar Lander* launched in 1999. It was supposed to study an area near the planet's south pole. But its landing gear malfunctioned, and it crashed. Also in 1999, the NASA *Mars Climate Orbiter* burned up in the sky over Mars due to a software error. Then the European Space Agency's *Schiaparelli* lander crashed in 2016 after spinning out of control on its way down to the surface.

SIGNS OF LIFE

Water isn't the only sign pointing toward the possibility of life on Mars. In 2018, scientists announced two exciting discoveries, one in the rocks and another in the air. *Curiosity* has been collecting sand and grinding up rocks since landing. It looks through the resulting bits and pieces for complex molecules that contain the chemical carbon. These are also called organic molecules. They are left behind when a living thing dies. But they can also form in other ways that don't involve living things. In 2018, *Curiosity* found organic molecules that could be the remains of ancient life. But NASA scientists did not know for certain.

The discovery in the air was similarly tantalizing. Previous Mars missions had found methane gas in the air. *Curiosity* went further, finding that the methane level rises and drops in a predictable cycle. This gas is interesting because on Earth, most methane comes from living things. However, the right combination of heat, water, and rocks can also produce methane without life. No one knows if alien microbes exist now or ever existed on Mars, but space scientists haven't yet ruled out the possibility. And *Curiosity* is still out there, gathering evidence.

SPACE EXPLORERS

Few people have had the opportunity to explore space in person. Around 550 humans have traveled into space.[9] Of these, just 24 have gone beyond Earth's orbit.[10] Yet over the past centuries, scientists have found many other ways to explore the cosmos. They have pointed telescopes at the stars. They have launched rockets, satellites, and space stations. They have sent probes, orbiters, landers, rovers, and other spacecraft to visit many parts of the solar system and beyond. They have used radio dishes, space telescopes, and other advanced equipment to study amazing structures from black holes to exploding stars called supernovae.

Every scientist involved in these missions is an explorer. Their feet may be on Earth, but their eyes and minds are out among the planets and stars. Steltzner says, "When we explore, when we are operating at the edges of our capability, we are fundamentally wondering about who we are as humans. . . . How great is our reach? What questions might we dare

"*Curiosity* is on a mission of discovery and exploration, and as a team we feel there are many more exciting discoveries ahead of us in the months and years to come."[11]

— John Grotzinger, Mars Science Laboratory mission, California Institute of Technology

COMPANY FOR *CURIOSITY*

Many more missions have followed *Curiosity* to the red planet. In 2018 the *InSight* lander arrived on Mars to study the planet's interior. The Mars 2020 rover is planned to launch in the summer of 2020. Part of its mission is to collect rock and soil samples that another mission will bring back to Earth to study. It will even carry a small drone helicopter—the first aircraft to ever take flight on another world. *ExoMars* is another Mars mission planned to depart in 2020. For this mission, the European Space Agency and Russian State Space Corporation are working together to operate a rover on the planet.

The scientists and engineers behind space missions explore the universe from control rooms on Earth.

ask and attempt to answer?"[12] Space explorers dare to ask big questions each day. They investigate how the universe began and how it may end. They study how stars and planets formed and changed, and they search for life on distant worlds. The things they discover help everyone better understand humanity's place in a vast, amazing universe.

CHAPTER TWO
THE RACE TO SPACE

A starry night sky fills the mind with wonder. Space scientists and explorers take that wonder to the next level. They seek answers about what is really out there. Throughout human history, civilizations have attempted to understand and explain the motions of Earth, as well as the sun, stars, planets, and other objects in the night sky. For example, the ancient Chinese, Mesopotamians, Greeks, and Maya all learned to predict eclipses.

The Greeks and Romans also had a theory that Earth was a perfect, stationary sphere and the sun and all of the planets traveled around it on perfectly circular paths. This isn't true, but in Western civilization it remained the dominant model of the universe for centuries. In 1543, Polish astronomer Nicolaus Copernicus published his radical new idea that Earth and all of the planets orbit the sun. This theory is known as heliocentrism.

People have been studying the night sky for thousands of years.

In the century that followed, scientists including Italian astronomer Galileo Galilei began to make more detailed observations of space using a new invention—the telescope. It revealed the sun, moon, and distant planets in far more detail than ever before. Galileo observed moons around Jupiter and spots on the sun. His observations of Venus proved that Earth and Venus both orbit the sun. The German astronomer Johannes Kepler's work supported heliocentrism, too. Kepler also proved that the planets' orbits are shaped like ellipses, not perfect circles.

> "I prove philosophically not only that the earth is round . . . but also that it is carried along among the stars."[1]
> —Johannes Kepler, astronomer, 1609

In the late 1600s, the British scientist Isaac Newton formulated his law of gravity and three laws of motion. His work helped cement the scientific understanding of the motion of planets and stars. Over time, larger, more powerful telescopes allowed scientists to discover new planets, moons, comets, asteroids, stars, and even distant galaxies. For centuries, astronomers made dramatic advances in space science without ever leaving the ground.

A POWERFUL ROCKET

Humanity could learn even more about space by going there. Launching anything into space is a huge challenge. Objects need to reach amazingly high speeds and altitudes to escape the atmosphere. Accelerating to such speeds requires incredible power. A rocket engine can provide that power.

Rocket technology has its roots in warfare. The medieval Chinese invented gunpowder and fireworks. In 1232, they launched simple missiles, which they called fire arrows, at Mongol invaders. Basic rocket technology later spread around the globe. But these early weapons weren't powerful enough to make it to space. Advances in design, construction, and fuel would be needed to escape Earth. The American engineer Robert H. Goddard made a major breakthrough when he launched the world's first liquid-fueled rocket in 1926. "It looked almost magical as it rose," he wrote in his diary.[2] The rocket only went 41 feet (13 m) into the air and reached 60 miles per hour (97 km/h), but Goddard believed that rocket technology would someday be able to carry equipment or even people into outer space.[3] He worked on ways to control a rocket's flight and to keep the materials it carried, known as the payload, safe.

Goddard's early rockets were the precursors to the machines that eventually carried humans to the moon and probes to the farthest reaches of the solar system.

During World War II (1939–1945), Germany developed long-range rockets for warfare. After the fighting ended, the United States and Soviet Union became embroiled in a Cold War. The two nations did not fight directly, but they competed in many ways. One way was technology. They each raced to develop better weapons, including better rockets. On October 4, 1957, the Soviet Union launched *Sputnik 1*. The size of a basketball, the silver craft was the first artificial object to reach outer space and orbit Earth. It had four long antennas that radioed a beeping sound back to Earth every time it completed an orbit. A new competition between the United States and the Soviet Union, known as the space race, had begun.

SURVIVING THE JOURNEY

Sending machines into orbit around Earth was an important milestone. But the real dream was much bigger. Human explorers longed to venture into space. Some thought this would be impossible. First, astronauts would have to survive the rocket ride into space. Once they arrived there, no one knew what might happen to their bodies. People speculated whether space travelers might not be able to breathe or digest their food. In 1957, the American inventor Lee de Forest said, "To place a man in a multi-stage rocket and project him into the controlling gravitational field of the moon. . . . I am bold enough to say that such a man-made voyage will never occur regardless of all future advances."[4]

But explorers don't let fear or difficulty hold them back. Scientists in the Soviet Union and the United States were determined to send human beings into space. Tests with animal passengers showed that it should be possible for people to survive the trip. Then, on April 12, 1961, a young pilot named Yuri Gagarin climbed aboard the Soviet spacecraft Vostok 1 and blasted off into history. He became the first human being to orbit the planet. The trip lasted 108 minutes.[5] "I saw for the first time the Earth's shape," he said after his return. "I could easily see the shores of continents, islands, great rivers, folds of the terrain, large bodies of water . . . the feelings which filled me I can express with one word: joy."[6]

ANIMALS IN SPACE

To test whether people might survive the trip to space, scientists first sent animals. The dog Laika became the first animal to make it into Earth's orbit in 1957 aboard the Soviet *Sputnik 2*. Sadly, she died a few hours into the trip. Two years later, the US space program launched a pair of monkeys into space and recovered them alive. The Soviet Union's *Sputnik 5*, launched in 1960, carried two dogs, a rabbit, 42 mice, two rats, and some fruit flies into orbit and then safely back home again.

THE MOON MISSIONS

One month later, President John F. Kennedy vowed that the United States would land a person on the moon and return the person safely to Earth before the end of the decade. He said, "No single space project in this period will be more impressive to mankind, or more important for the long-range exploration of space; and none will be so difficult or expensive to accomplish."[7] A moon mission became the number one priority for NASA. Getting there required a lot of preparation. The Mercury missions, launched between 1961 and 1963, put the first American astronauts into space. The Gemini missions of 1965 and 1966 tested new flight systems and equipment and sent astronauts outside their craft on

HUMAN COMPUTERS

All of the people NASA sent into space during the 1960s and 1970s were white men. Behind the scenes at NASA, though, women of color worked as computers. They did the mathematical calculations needed to design rockets and plan space missions. The 2016 movie *Hidden Figures* explored the experiences of three African American women who did this important work—Katherine Johnson, Dorothy Vaughan, and Mary Jackson.

space walks. Each mission built on lessons learned on past missions. They paved the way for Apollo, the lunar exploration program.

On the Apollo 8 mission in 1968, humans orbited the moon for the first time. And on July 20, 1969, 600 million people around the world watched on television, transfixed, as astronaut Neil Armstrong stepped out of a lunar lander and onto the surface of the moon at the climax of the Apollo 11 mission.[8] "That's one small step for [a] man, one giant leap for mankind," Armstrong said.[9] The astronauts planted a flag and a plaque on the moon. The plaque says, "Here men from the planet Earth first set foot on the moon—July 1969 A.D.—We came in peace for all mankind."[10] Five more Apollo missions landed successfully on the moon. The spacefarers on these missions collected samples, took photographs, and rode in a carlike vehicle called the lunar rover.

SPACE SHUTTLES AND STATIONS

Just 12 humans have set foot on the moon. The last of these missions happened in 1972. But human spaceflight continued with the development of the space shuttle, a reusable vehicle that carried astronauts into space and landed like an airplane back on Earth. NASA launched space shuttle missions from 1981 through 2011. Part of the shuttle's job was to

By 2019, the Apollo astronauts were still the only humans to ever explore another world on foot.

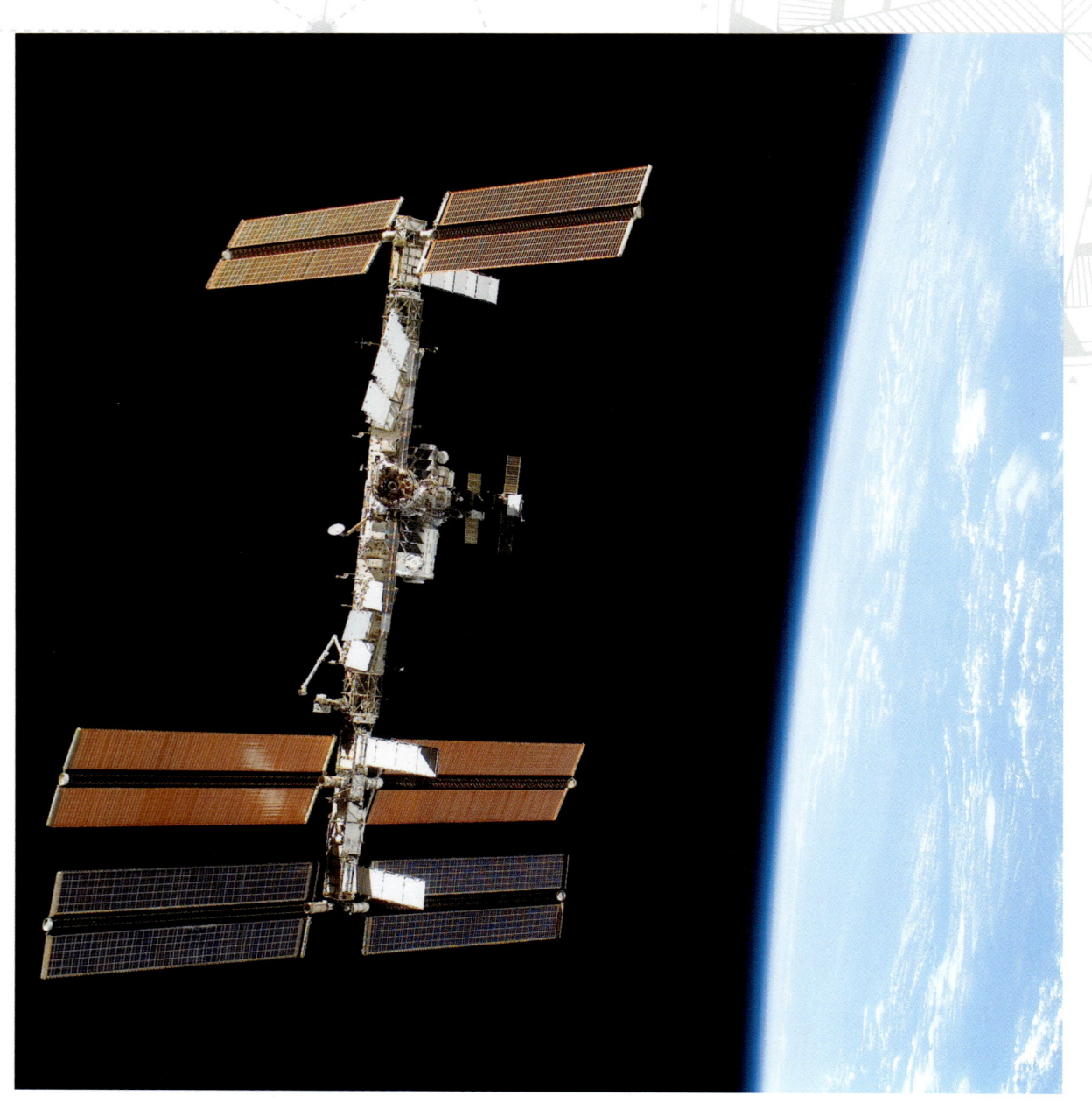

The ISS is one of the most expensive and complex engineering projects in world history.

help build the International Space Station (ISS). The ISS is operated by the United States and Russia, with contributions from several other nations. In 1991, the Soviet Union dissolved, splitting into several nations, the largest of which is Russia. Since then, the United States and Russia have cooperated in space.

Construction on the station began in 1998. The European Space Agency (ESA), Canadian Space Agency, and Japanese Aerospace Exploration Agency played key roles in the program. The main construction was completed in 2011, and the station is expected to remain operational through 2024. When people gaze up into the night sky, they can see the ISS as a slowly moving light. Astronauts aboard the station conduct experiments, study the effects of space on the human body, and expand our knowledge about living and working in space. This knowledge will be crucial for longer voyages of exploration to Mars and other planets.

"As I take man's last step from the surface, back home for some time to come—but we believe not too long into the future—I'd like to just [say] what I believe history will record. That America's challenge of today has forged man's destiny of tomorrow. And, as we leave the Moon at Taurus–Littrow, we leave as we came and, God willing, as we shall return, with peace and hope for all mankind. Godspeed the crew of Apollo 17."[11]

—*Eugene Cernan, the last Apollo astronaut to leave the lunar surface, 1972*

TRAGEDY STRIKES

Astronauts ride to space at the tips of massive rockets filled with explosive fuel. If something goes wrong, the results may be tragic. One such disaster happened in 1986, when the space shuttle *Challenger* exploded in midair. The seven crew members died. Another disaster occurred in 2003. The space shuttle *Columbia* suffered damage to its heat shield during launch. When it returned to Earth 15 days later, the damage resulted in the shuttle burning up in the atmosphere. All seven crew members died. Despite the extreme risks, many people still volunteer to explore space.

CHAPTER THREE
LIFE AS AN ASTRONAUT

Three people lie on their backs, strapped into their seats in a tiny capsule perched on top of three powerful engines, two massive rocket boosters, and an even larger tank of rocket fuel. The countdown begins: "Ten, nine, eight, seven, six."[1] The engines roar to life. At zero, the rocket explodes into the sky.

Astronaut Chris Hadfield, a veteran of two space shuttle flights, describes the experience like this: "You are in the grip of something that is vastly more powerful than yourself. It's shaking you so hard you can't focus on the instruments in front of you. It's like you're in the jaws of some enormous dog and there's a foot in the small of your back pushing you into space, accelerating wildly straight up."[2] This intense experience lasts a few minutes before astronauts reach space, when the feeling is replaced by calm and weightlessness.

Launching on a rocket can be a jarring, violent experience. Still, astronauts willingly endure it for the opportunity to visit space.

At any given time, just three to six people live and work on the ISS. They come from countries around the world. Yet thousands of people apply for astronaut positions each time NASA asks for new recruits. The job is unique and highly desirable for those who want to go faster, higher, and farther than all other explorers. In 2007, engineer and NASA astronaut Leland Melvin took his first trip to space. Before beginning his assigned tasks, he looked down at Earth. "I push off and float over to the window, and I can see the Caribbean. And I need new definitions of blue to describe the colors that I see. Azure, indigo, navy blue, medium navy blue, turquoise don't do any justice to what I see with my eyes."[3] Many astronauts have described similar moments of wonder and awe. But the job is about much more than amazing views. Astronauts perform difficult, technical, scientific work every day.

"I feel like a little kid, like a sorcerer, like the luckiest person alive. I am in space, weightless, and getting here only took 8 minutes and 42 seconds. Give or take a few thousand days of training."[4]

—Chris Hadfield, astronaut

FROM THE NFL TO OUTER SPACE

Leland Melvin always wanted to be a professional sports player. In 1986, the Detroit Lions drafted him. But after an injury during training camp, he had to find a new dream. He later earned a degree in materials science engineering and went to work for NASA. Eventually, he became an astronaut. He says his football experience helped prepare him for space. "The thing about football is that when it gets really noisy in the stadium a quarterback and a wide receiver have to have this nonverbal ability to communicate as the coverage changes," he says. "You have to have that same ability to communicate when you're on the flight deck."[5]

FRAGILE BODIES AND MINDS

An important subject that astronauts study is what happens to the human body as people spend time in space. Conditions in space, even inside a spacecraft,

Leland Melvin flew two missions into space.

Astronauts have occupied the ISS continuously since November 2, 2000.

differ significantly from those on Earth. On the ground, a thick blanket of atmosphere protects people against radiation from the sun and other sources. In space, the only protection comes from the thin metal skin of the spacecraft. This can result in higher doses of dangerous radiation. Astronauts are also freed from the effects of gravity. Though Earth's gravity still acts upon them in space, they experience weightlessness because the ISS is traveling fast enough to remain in a stable orbit. Essentially, they are in a constant state of free-fall as they circle the planet. This condition is known as microgravity.

To sleep, astronauts attach a sleeping bag to a wall and crawl inside. Going to the bathroom involves tubes and vacuums to make sure no waste escapes into the air. But weightlessness is also great fun. Bouncing off the walls and turning somersaults takes almost no effort. Mike Massimo, a NASA astronaut, loved eating in space. He says, "You have to be careful,

The International Space Station orbits Earth every 90 minutes. That means the people on board get to see 16 sunrises and 16 sunsets every single day![6]

SAVE THAT PEE

Water is a precious resource in outer space. It's expensive to send fresh water to the ISS, so the station has a system that collects and cleans astronauts' urine. It becomes clean enough to drink again. That may seem gross, but to space scientists, it's amazing. "Where other people see urine, I see a vital resource that is going to help astronauts survive," says water recovery engineer Dean Muirhead.[7] For a long space flight, such as a trip to Mars, astronauts would need to survive without any new supplies arriving from Earth.

because everything floats, but that's the fun part: Popping M&Ms in the air and going after them and chomping them like Pac-Man."[8]

Floating is fun, but bones and muscles weaken rapidly when they don't have to work to support the weight of the body. Astronauts must exercise for two hours every day just to avoid losing too much strength. In addition, all the fluids in the body float upward, which gives many astronauts headaches and nausea for a few days. Astronauts experience psychological challenges, too. They live in cramped quarters and face the mental stress of separation from family and friends.

On a space mission that lasts a year or more, as journeys to Mars and other planets would, astronauts' bodies and minds would go through many extreme changes. To help researchers better understand these changes, astronaut Scott Kelly volunteered to become a test subject. He spent 340 days from 2015 to 2016—almost an entire year—living on the ISS. Meanwhile, his identical twin brother, astronaut Mark Kelly, stayed behind on Earth. The two brothers both went through regular tests. Researchers took blood, urine, and microbial samples. The brothers played games intended to test memory and reaction time. Researchers compared the results against each other to see how life in space and the eventual return to Earth changed Scott's body.

Some of the changes seem alarming. While in space, Scott's immune system ramped up its activity. In addition, the population of microbes living in his gut changed drastically. He also experienced more genetic mutations than Mark, most likely because he was exposed to more radiation. Genetic structures called telomeres also changed. These sections of DNA sit at the ends of chromosomes and act sort of like the plastic tips on shoelaces, protecting genetic information from getting scrambled. Normally, telomeres get shorter as people get older or go through stress. But Scott's got longer. "We don't know yet if these [genetic] changes are good or bad," says Christopher Mason, a geneticist at Weill Cornell Medicine in New York.[9] The telomeres, microbes, and immune system all quickly returned to normal after Scott came home. Studying the changes will help scientists find better ways to keep people safe during and after long space missions.

AMAZING MICROGRAVITY

Space-based research also benefits people on Earth. Microgravity affects all sorts of living things and materials in unusual ways. Some of those effects are very useful. Scientists on Earth who specialize in medical research, materials science, engineering, and many other areas regularly design experiments and launch them to the space station. Then astronauts help run and monitor the experiments.

For example, to create drugs that will cure a disease, researchers have to figure out the intricate three-dimensional structures of molecules called

SPACE AGE INVENTIONS

While developing the technology necessary to send people to space and keep them happy and healthy there, scientists and engineers have come up with ideas and inventions that are now a part of daily life. Lightweight survival blankets, memory foam mattresses, scratch-resistant glass, shock-absorbing running shoes, and many other items were all developed by engineers working to solve problems related to space travel.

Astronauts have grown plants in space, testing out techniques that could be helpful in future long-duration missions.

proteins. One way to do this involves growing crystals of the protein being studied. In microgravity, these crystals grow larger and more perfectly than they do on Earth. Experiments in outer space may lead to new treatments for a number of different diseases, including hepatitis C, Huntington's disease, cystic fibrosis, and some cancers.

Microgravity also changes the way metals mix to form materials called alloys. A device called the electromagnetic levitator (EML) helps astronauts melt metals, mix them together, and allow them to cool. Metals that might rise or sink on Earth end up mixed much more evenly. These experiments could help scientists design new materials, including better turbine blades for power plants.

BUILDING AND MAINTENANCE

Astronauts have many jobs in addition to research. They must also keep the space station in working order. They conduct routine safety checks, inspect or clean equipment, and perform repairs as needed. They also install new systems or upgrades. In 2018, an unknown object—likely a tiny space rock called a micrometeoroid—punched into the side of one of the laboratories on the space station. Air began leaking out very slowly. Cosmonaut Sergey Prokopyev placed a piece of gauze soaked in powerful glue over the hole. That solved the problem. The leak was small enough that the astronauts were never in real danger. But small problems must be addressed quickly before they get worse.

Some maintenance and repair jobs require working outside the space station. For these tasks, astronauts go on space walks. In March 2019, Anne McClain and Nick Hague put on bulky space suits and helmets and stepped out into space. Each space suit is like a miniature spacecraft. The suits provide air for breathing and keep human bodies at a safe temperature. A tether line kept the astronauts attached to the space station so they didn't float away. The pair replaced batteries in the solar panels that provide electricity to the space station.

A STUNNING VIEW

The experience of seeing Earth from afar only gets better on space walks. Chris Hadfield says it's difficult to describe the feeling of being alone in the universe, far above Earth. "It is like coming around a

Astronauts on the ISS have gained valuable experience doing maintenance and repairs on space walks. Such tasks will be crucial for future space exploration.

corner and seeing the most magnificent sunset of your life, from one horizon to the other where it looks like the whole sky is on fire and there are all those colors. . . . You just want to open your eyes wide and try to look around at the image, and just try and soak it up. It's like that all the time."[10]

> "It can be crowded, noisy, and occasionally uncomfortable. Space travel—at least the way we do it today—isn't glamorous. But you can't beat the view!"[11]
>
> —*Marsha Ivins, astronaut*

Astronauts are space travelers who go on an adventure like no other when they depart their home planet. They risk their lives and often go through extreme discomfort and stress. Yet they love their jobs. They are helping to pave the way for future explorers who may travel to Mars or even more distant frontiers.

Microgravity allows for unique portraits of the astronauts on the ISS.

SCIENCE CONNECTION
MICROGRAVITY

"Every day is a good day when you're floating," says astronaut Anne McClain, who arrived on the ISS in 2018. "Your whole life you spend walking around Earth and then all of a sudden you get to fly like you've dreamed of."[12] Why do astronauts float? The answer involves the science of gravity and motion.

Gravity is a force that pulls objects toward each other. The more massive an object is, the harder it pulls. Even when they are standing still, people's muscles and bones are working to support them against Earth's gravity. The lower the pull of gravity, the less a person weighs, even if his or her body size stays exactly the same. The pull of Earth's gravity does weaken with distance. But the ISS isn't far enough away for that to explain the way astronauts float. A person who weighs 100 pounds (45 kg) on Earth would weigh 90 pounds (41 kg) if he or she were standing stationary at the top of a ladder that reached all the way to the space station.[13]

The key to microgravity is that the space station isn't standing still. It's moving extremely quickly. It zips around Earth at roughly 5 miles (8 km) per second.[14] Despite this speed, Earth's gravity prevents the station from zooming off into deep space. It also doesn't get pulled all the way to the ground because it's moving so fast. The station is basically constantly falling and constantly missing Earth. It ends up moving in a circular path, or an orbit, around the planet. Everything inside the space station, including the people, is orbiting Earth at the same speed. Therefore, nothing falls into anything else. It all floats.

CHAPTER FOUR
MYSTERIOUS MOONS

Human explorers haven't yet ventured beyond Earth's moon. But robotic spacecraft have gone much farther. They have visited many places in Earth's cosmic neighborhood, peering at the planets and their moons up close. These explorers have made many stunning discoveries—especially about the solar system's dozens of moons.

For a long time, most scientists thought all moons were similar chunks of barren rock. But then the spacecraft *Voyager 1* flew past Io, one of Jupiter's moons, on March 4, 1979. The images the spacecraft sent back stunned scientists. "It's got volcanoes and sulfur, it looks like a pizza and has some sort of bizarre surface," says astronomer Joseph Burns.[1] This was the first time volcanoes had been found anywhere beyond Earth. Over the decades, as spacecraft visited more moons, the surprises piled up. Each moon was its own unique world.

The *Voyager* spacecraft and later probes have made stunning discoveries about the moons of the outer solar system.

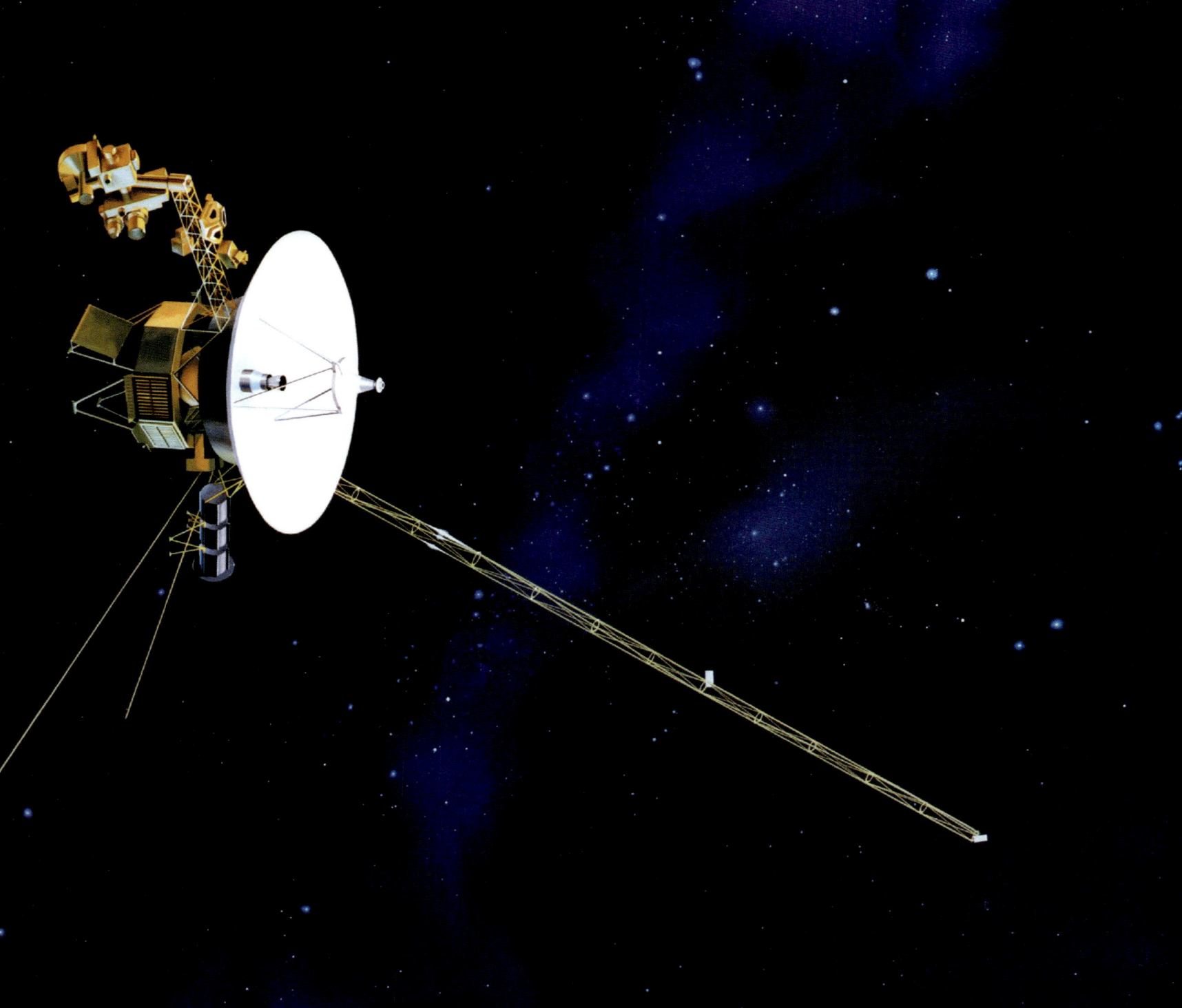

THE PALE BLUE DOT

In 1990, the spacecraft *Voyager 1* snapped a photo of planet Earth from afar. *Voyager 1* was at the edge of the solar system, more than 3.7 billion miles (6 billion km) from the sun. At this distance, Earth was just a tiny speck. Carl Sagan, an astronomer who worked on the *Voyager* team, said this about the image: "Look again at that dot. That's here. That's home. That's us. On it everyone you love, everyone you know, everyone you ever heard of, every human being who ever was, lived out their lives. . . . To me, it underscores our responsibility to deal more kindly with one another, and to preserve and cherish the pale blue dot, the only home we've ever known."[4]

THE BURIED OCEAN

The *Voyager 2* spacecraft, which arrived at Jupiter a few months after *Voyager 1*, investigated Europa, another of Jupiter's moons. Its images showed a world covered in ice. Scientists suspected liquid water might lurk beneath the surface. A later mission explored this question further. In the 1990s, the *Galileo* spacecraft studied the planet. It discovered that as Europa orbits Jupiter, the moon somehow causes changes in the huge gas planet's magnetic field. The best explanation for these changes is a vast, salty ocean under the moon's outer shell. "We're almost certain one is there," says scientist Cynthia Phillips.[2]

The Hubble Space Telescope, which studies the universe from its orbit around Earth, has added to the evidence of a liquid ocean. Pictures taken in 2012, 2014, and 2016 seem to show plumes coming off the same location on Europa's surface. These are most likely jets of water vapor erupting through the ice from a liquid ocean beneath. As on Mars, the discovery of liquid water means that life may also exist. "Europa's ocean is considered to be one of the most promising places that could potentially harbor life in the solar system," says NASA official Geoff Yoder.[3]

ICE GEYSERS

Another moon that may harbor life orbits Saturn. Enceladus is a small, bright moon. In fact, it's the most reflective of all the known planets and moons. That's because it's completely covered in white ice. The spacecraft *Cassini* arrived at Saturn in 2004, and it soon noticed that something strange was going on with this moon. Scientists had the spacecraft take a closer look. In 2005, the *Cassini* team discovered huge plumes of vapor and ice particles erupting from near the moon's south pole. The expelled material feeds one of Saturn's most distant rings. Linda Spilker, the *Cassini* Project Scientist at the Jet Propulsion Laboratory, was astonished. "Enceladus . . . should have frozen solid long ago yet today it lofts icy particles and gas into space."[5]

> ### WRONG WAY!
> Jupiter has a lot of moons—79 and counting. In 2017, scientists discovered 12 new moons. One of them was especially strange, orbiting in the opposite direction of nearby moons. Like a car driving down the wrong side of the road, it's at risk of a crash. "This is an unstable situation," says astronomer Scott S. Sheppard of the Carnegie Institution for Science, who led the team that discovered the moons. "Head-on collisions would quickly break apart and grind the objects down to dust."[6] Until that happens, the new moon has the name Valetudo.

As on Europa, these plumes likely come from a vast, salty liquid ocean hidden beneath the icy surface. Scientists have not yet seen Europa's plumes up close. But in 2008, *Cassini* actually flew through the Enceladus plume and collected information about its contents. Scientists have spent years analyzing this data and all the other information *Cassini* gathered about the plume and the moon's surface. In 2015, scientists realized that hydrothermal activity, or the movement of heated water, probably explains the plumes. In Earth's oceans, hot magma beneath Earth's crust heats some water and forces it up into the

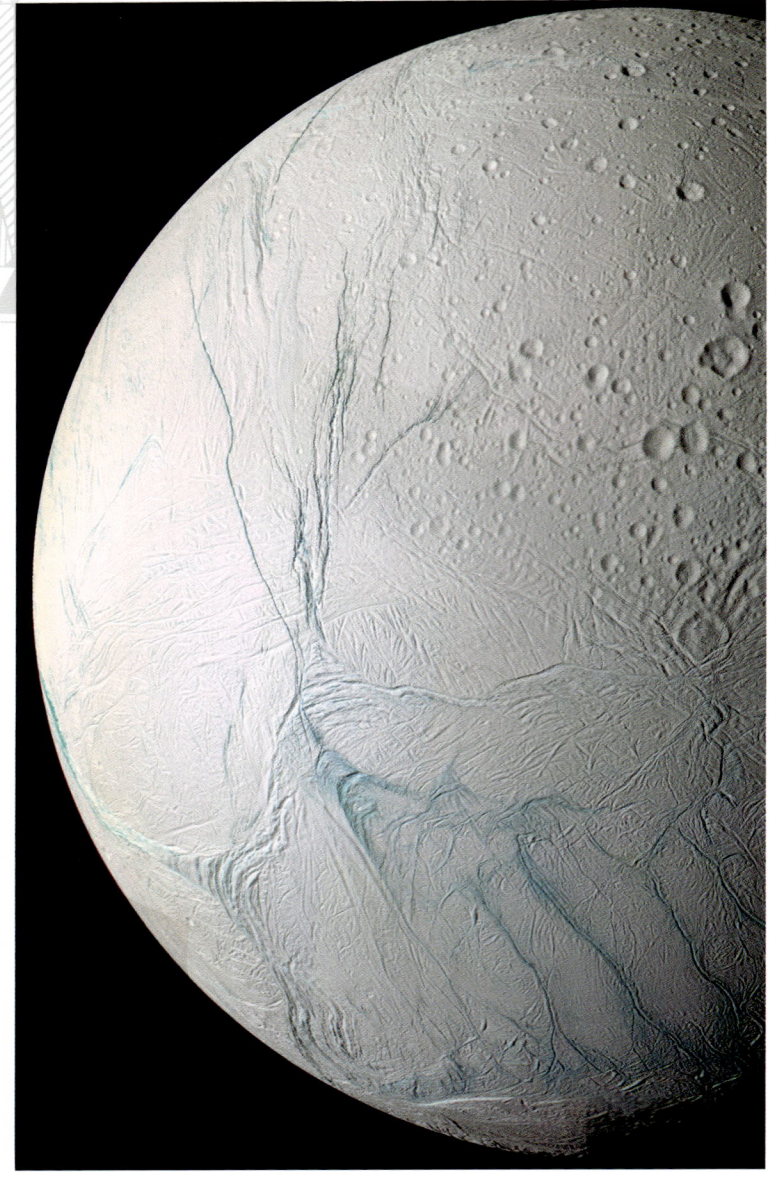

Scientists have found that Enceladus's icy surface likely hides a salty ocean.

ocean through cracks in the seafloor, called hydrothermal vents. Many living things flourish around these vents. Some scientists believe life may have first emerged in a place like this. If Enceladus has similar vents, perhaps life appeared there, too.

The ice grains *Cassini* encountered during its pass through the plume were large organic molecules, including propane and benzene. Carolyn Porco, the imaging team leader for the *Cassini* mission, says that thanks to this new research, scientists "are much more confident now than we were two years ago that we might indeed have on this moon, under the south pole, an environment or a zone that is hospitable to living organisms."[7] Organic molecules are the ingredients life requires to form. It takes a series of chemical reactions to produce large, complex organic molecules. These reactions happen in living things, though they may also occur during hydrothermal or geologic activity.

Cassini's mission ended in 2017. To find out whether life actually exists on Europa, Enceladus, or elsewhere in the solar system, space scientists hope to send new spacecraft to probe or even land on these distant worlds. NASA plans to launch such a mission, called the *Europa Clipper*, in the 2020s.

> "I invite you to imagine the day when we might journey to the Saturnine system, and visit the Enceladus interplanetary geyser park, just because we can."[8]
> —*Carolyn Porco,* Cassini *imaging team leader*

YELLOW LAKES

One distant moon landing has already taken place. Saturn's largest moon, Titan, has intrigued scientists ever since *Voyager 1* flew past it in 1980. The spacecraft showed that the moon has a thick atmosphere filled with orange clouds containing methane, ethane, and other organic compounds. On Earth, methane and ethane are invisible gases. But Titan is so cold that these substances may turn into liquid and fall as rain. They may even flow across the surface as rivers, lakes, and seas.

In 2005, *Cassini* sent a small lander probe it carried, *Huygens*, down to the surface of Titan. The probe took measurements and pictures during its two-hour descent to the surface. No one knew what *Huygens* might land on or in, so it was designed to float. It landed on a frozen, dry plain littered with rocks. It transmitted data for 72 minutes, only

The sturdy *Huygens* lander was engineered to survive a seven-year journey through space aboard *Cassini* before descending to the surface of Titan.

stopping when its batteries ran out. *Huygens* was the first probe to land on any planet or moon beyond the asteroid belt. Thanks to the probe and *Cassini*'s investigations, scientists now know that Titan does have lakes of liquid methane. "Methane plays the role on Titan, with clouds and rain, that water plays on Earth," says Spilker.[9] The moon also has vast dunes made up of an unknown material containing hydrogen and carbon.

BEYOND NEPTUNE

The solar system has eight planets. Pluto, once considered a planet, was reclassified as a dwarf planet in 2006. Astronomers had a good reason for this decision. They had begun discovering many more small, rocky worlds past Neptune. Scientists felt that either they should all be planets, or none of them should be. In the end, astronomers decided to call them all dwarf planets.

The region with all the new dwarf planets is called the Kuiper Belt, named for astronomer Gerard Kuiper. Scientists have detected more than 2,000 objects there. One of them, Eris, is a rocky world with its own moon. It's slightly larger than Pluto but follows a very stretched-out orbital path that sends it extremely far away from the sun. Michael Brown and his team at the California Institute of Technology discovered Eris in 2005. Astronomers think that hundreds of thousands of objects wider than 60 miles (100 km) across remain to be discovered in the Kuiper Belt.

The Oort cloud is a region that lies past the Kuiper Belt. It's another band of objects and is the likely birthplace of most comets. In 2003, a few years before finding Eris, Brown and his team found an Oort cloud object that they called Sedna, named in honor of an Inuit goddess of the sea. The dwarf planet is about half the diameter of Earth's moon. It follows an extremely broad path around the sun, taking 11,000 years to complete one orbit.

THE HEART AND THE SNOWMAN

NASA's *New Horizons* spacecraft launched in 2006 for a very long journey. It was heading for Pluto and the Kuiper Belt. In 2015, after almost a decade of travel, the spacecraft finally arrived at Pluto. It sent back a cosmic valentine. The dwarf planet, it turns out, has a feature on its surface that resembles a giant heart. A few years later, in early 2019, *New Horizons* checked out a more distant Kuiper Belt object called Ultima Thule. It looks a bit like a snowman, with two spheres stuck together. *New Horizons* continues to fly through the Kuiper Belt. Scientists are excited to see what other strange things it may uncover.

PLANET 9

Something is affecting the orbits of objects in the Kuiper Belt. One explanation for the unusual orbits is another planet. If it exists, it would probably be around five to ten times larger than Earth. Astronomers call it Planet 9. But they have not yet managed to detect it. "This is a very dim object in a very big sky. Since we don't know exactly where it is, you have to survey the whole sky, or at least large portions of it, in order to find the planet," says Fred Adams of the University of Michigan.[12] Many astronomers believe that the planet exists and will soon be located.

Brown describes what it's like to discover something like Eris or Sedna: "You go through all this data and there's nothing there, nothing there, nothing there, and then suddenly there's something that no one has ever seen before except for you. It's always a moment of excitement."[10] Linda Spilker agrees. She enjoys her job as a space scientist because it gives her "the chance to be an explorer: to go places and see things for the very first time."[11] Using telescopes and spacecraft, scientists such as Brown and Spilker continue to discover amazing things about the planets, dwarf planets, and moons of the solar system.

Pluto, *right*, and its moon, Charon, *left*, are two of many thousands of objects drifting through the Kuiper Belt.

The US portion of the Deep Space Network includes several huge, dish-shaped antennas located in California's Mojave Desert.

SCIENCE CONNECTION
TALKING TO SPACECRAFT

Operating a rover such as *Curiosity* or a spacecraft such as *Cassini* is sort of like driving a remote-control car. Scientists on Earth send instructions in the form of radio waves that tell the rover or spacecraft where to go and what to do.

One tricky part about space communication is that those waves must travel very long distances. Even though radio waves move at the speed of light, it takes between four and 24 minutes for a signal from Earth to reach Mars, depending on how far away the two planets are at the time. And reaching Saturn takes around 90 minutes. Getting a signal out to the Kuiper Belt takes even longer.

The scientists operating distant spacecraft plan out long chunks of instructions that tell the spacecraft what to do every minute of every day until the next set of instructions arrives. Most are also programmed so that if they end up in a tricky situation, they can stay safe. For example, *Cassini* was programmed not to turn toward the sun, because this could damage its cameras. *Curiosity* has hazard avoidance software that stops the rover every ten seconds to check its surroundings for possible dangers, such as large rocks or holes.

Spacecraft also send information back to Earth. The Deep Space Network picks up the information. This network has large antennas spaced out around the globe in Australia, Spain, and the United States. Each site has several antennas, including one 230 feet (70 m) in diameter.[13] Thanks to the sites' spacing, scientists can pick up the signal no matter which way Earth is facing.

CHAPTER FIVE

CATCHING COMETS AND ASTEROIDS

Planets, dwarf planets, and moons aren't the only things in the solar system that scientists and their spacecraft explore. They also investigate comets and asteroids. Both are like time capsules left over from the formation of the solar system. Billions of years ago, comet impacts likely delivered ice and organic molecules to the early Earth. "By going and studying a comet at the present day you can actually look back in time at what the Earth was formed from," explains Ian Wright, a planetary scientist at the Open University in the United Kingdom.[1] A comet or an asteroid can teach scientists a lot about the early history of Earth and the solar system. But scientists also explore these space rocks because it's an exciting challenge. "We want to go to a comet

Up-close encounters with comets have helped scientists better understand these mysterious objects from deep space.

COMETS VS. ASTEROIDS

Comets and asteroids are much smaller than planets. Comets formed far from the sun, and they contain ice, dust, and rocky material. Some comets travel on orbits that regularly swing close to the sun. During this part of the orbit, some of a comet's ice vaporizes to form a tail. Asteroids formed closer to the sun and are made mostly of metal and rock, so they don't usually have tails. Since they are so small, they don't have much gravity. They often have irregular shapes. Most asteroids in the solar system orbit in the asteroid belt between Mars and Jupiter.

"Comets . . . are the travelers. They come from a distant place in space, and . . . we think they represent pristine, unchanged remnants of the distant past. They come to us as ambassadors, if you like, from a different time."[4]

—Claudia Alexander, *Rosetta project scientist, Jet Propulsion Laboratory*

because it's there," says Wright. "It's an object in our astronomical backyard and we want to know what it looks like and what it's made of."[2]

Wright got his wish to explore a comet. But he didn't go himself. He was part of the team that worked on the *Rosetta* mission, run by the ESA. The spacecraft *Rosetta* launched in 2004, carrying the small lander *Philae*. For ten years, the spacecraft spiraled through the solar system, before finally meeting up with the comet 67P/Churyumov-Gerasimenko in 2014. This comet's path takes it from just outside of Jupiter's orbit to in between the orbits of Earth and Mars, orbiting the sun once every six and a half years.

BOUNCING ON A COMET

Getting to comet 67P was the easy part of the mission. Next, the team planned to land *Philae* on the comet's surface. "We're trying to do something that has never been tried before in the history of humankind," Wright said.[3] Landing on a comet is tricky because the frozen space rock is so small. And small objects

The sophisticated *Philae* lander carried out the first-ever landing on a comet.

have low gravity. That means things don't stay stuck to the surface very easily. "There is significant risk of bouncing off the comet," explains Tom Marsh of the University of Warwick.[5] If the lander were to bounce upward with too much speed, it might float off into space, never to return. To help prevent this fate, *Philae* was equipped with two harpoons it was supposed to fire. These would anchor it to the surface. It also had drills in its tripod legs.

Finally, the big day came. On November 12, 2014, *Rosetta* released *Philae*. It slowly floated down toward the comet about as gently as a sheet of paper falling toward the floor. But then there was a problem. The harpoons didn't fire. Instead of landing and sticking, *Philae* bounced—twice! The first bounce lasted two hours, but thankfully *Philae* didn't fly off into space. It came down again and bounced once more, but only for a few minutes this time. Scientists didn't

***Philae* provided the closest-ever images of a comet. One of the lander's legs is visible in its photo from comet 67P's surface.**

know where exactly the lander was located. But they knew it was safe. For three days, it transmitted data about the surface of the comet. Unfortunately, its solar panels were shaded, so it couldn't charge its batteries from where it landed. *Philae* went silent.

Rosetta remained in orbit around the comet, studying how the tiny hunk of rock and ice changed as it neared the sun. Briefly, in 2015, when the comet was as close as it gets to the sun, *Philae*'s solar panels received enough light to wake up the lander. *Philae* sent a few bursts of data to *Rosetta* before going silent again. Then, in 2016, *Rosetta* finally found the lander on its side, wedged into a crack on the comet's surface. At the time, the comet was heading away from the sun, and the distance meant it was getting harder for *Rosetta* to keep its solar-powered batteries charged. So the team decided to land *Rosetta* on the comet, too. After that, it would cease communications. "We're trying to squeeze as many observations in as possible before we run out of solar power," said Matt Taylor, *Rosetta* project scientist at ESA.[6] But the mission didn't end when *Rosetta* went silent in late 2016. Scientists will spend years analyzing the data the mission collected.

ASTEROID ADVENTURES

Spacecraft have enabled scientists to explore asteroids, too. On February 14, 2000, NASA's *NEAR Shoemaker* spacecraft arrived at Eros, one of the largest asteroids in the solar system. Eros is named for a Greek god of love, making the Valentine's Day arrival appropriate. For a year, the spacecraft studied Eros from orbit. Then, in 2001, *NEAR Shoemaker* landed safely on Eros and gathered data for several days before ceasing communications.

In 2005, the Japanese Aerospace Exploration Agency (JAXA) spacecraft *Hayabusa*, whose name means "peregrine falcon," collected a sample from an asteroid named Itokawa. The plan was to fire a small pellet into the asteroid's surface and then suck up the dust and debris released from the impact. The pellet system didn't work as planned, but scientists determined that when the spacecraft contacted the asteroid, grains of material from Itokawa were likely knocked into the sampling device. They confirmed the discovery when *Hayabusa* returned home carrying 1,500 grains of dust from the asteroid.[7]

In 2011, NASA's *Dawn* spacecraft arrived in the asteroid belt. It visited egg-shaped Vesta first, then Ceres, which is the largest known asteroid. Ceres is so big that many consider it to be a dwarf planet, like Pluto. *Dawn* discovered that Ceres has volcanoes and the remnants of a frozen ocean. Those were surprising finds on such a tiny world. *Dawn* ran out of fuel and ended its mission in 2018.

Regolith is the word for a powdery material that covers most asteroids. It's likely dust left over from meteorite impacts.

IMPACT!

A large asteroid hit Earth 65 million years ago, wiping out most of the dinosaurs. At some point in the future, another large rock will hit Earth. Scientists plan to be ready. "The difference between the dinosaurs and us is that we have a space program," says astronomer Phil Plait.[8] Astronomers keep an eye on asteroids and will warn people in advance if one is on a collision course. Then NASA and other space programs can work together to respond. For example, a spacecraft could potentially knock an asteroid away from Earth, altering its course so that it misses the planet.

ASTEROID MINING

Missions to comets and asteroids help scientists learn more about these celestial bodies, and they also help lay the groundwork for future human missions. Asteroids could be useful for long-duration space travel. If people ever set up colonies on Mars or the moon, they'll need resources. Experts think the best way to get water, fuel, and other resources in deep space is to mine asteroids.

The asteroid Ryugu got its name from a Japanese folk tale. In the story, Ryugu-jo, which means "dragon's palace," was a magical palace in the ocean.

Just a few weeks before *Rosetta* landed in its final resting place on comet 67P, another spacecraft launched from Cape Canaveral Air Force Station in Florida. NASA's *OSIRIS-REx* was heading toward the asteroid Bennu. Although most asteroids orbit in a vast belt between Mars and Jupiter, this one orbits closer to home, between Earth and Mars. In early December 2018, after more than two years in space, *OSIRIS-REx* arrived at Bennu. The team planned to study the asteroid for two years. They also sought to collect a sample from the surface of the asteroid and then return it to Earth. *OSIRIS-REx* isn't supposed to land on Bennu. Instead, a long collection arm will extend and punch the surface, kicking up material into a storage container.

BRINGING SAMPLES HOME

Meanwhile, JAXA's *Hayabusa2* is already making history, exploring the asteroid Ryugu. As with Bennu, this asteroid's orbit brings it between Earth and Mars. In September 2018, *Hayabusa2* dropped two tiny robots, MINERVA-II1A and MINERVA-II1B, onto Ryugu's surface. Each was about the diameter of a small dinner plate. These were the first rovers

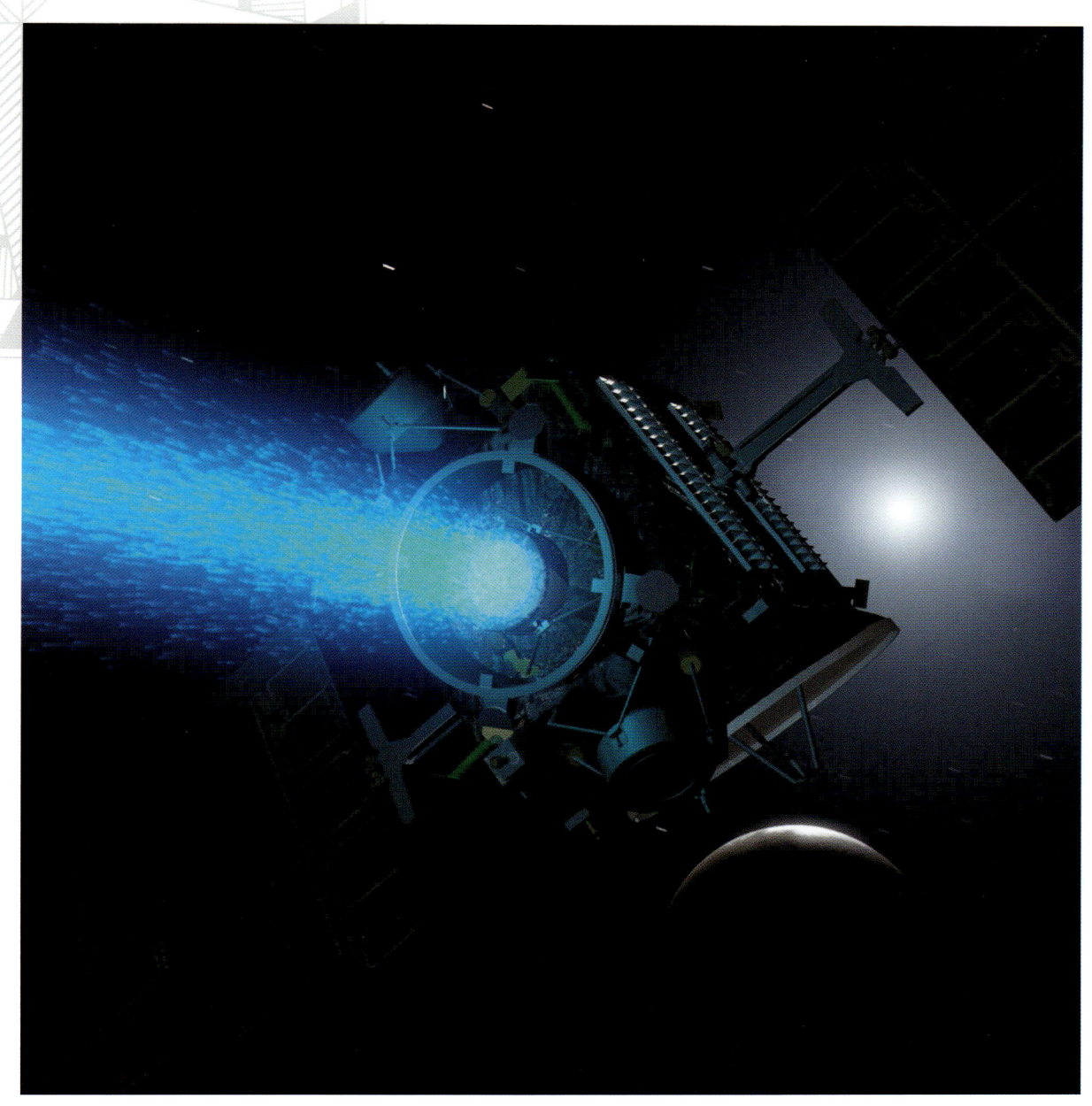

Dawn's ultra-efficient ion engine propelled it to the distant asteroid belt.

Hayabusa2 launched from Japan in 2014.

deployed on an asteroid. The original *Hayabusa* had also carried a MINERVA robot, but that robot never managed to land. The rovers have an unusual way of getting around. Instead of rolling, they hop, giving them their nickname, "hoppers." Gravity is so weak on Ryugu that a wheeled vehicle would just float upward instead of remaining on the surface.

In early 2019, *Hayabusa2* successfully executed the pellet-firing plan that *Hayabusa* had been unable to complete. Next, the spacecraft shot a bomb at the asteroid. It then quickly retreated to the other side of the space rock to avoid being pelted with debris from the explosion. The bomb made a crater. The team plans to collect a sample from inside the crater. The inside of the asteroid has been protected from sunlight and other forms of radiation for possibly billions of years. That makes it an exciting material for scientists to study. It's an even more pristine time capsule than the asteroid's surface.

Takashi Kubota, a JAXA spokesperson, is thrilled with the success of the mission. "I felt awed by what we had achieved," he says. "This is just a real charm of deep space exploration."[9] Missions to comets and asteroids probe the history of the solar system. And they also reveal the boldness and creativity of space scientists. To better understand the solar system, they are willing to go places no one has been and try techniques that have never been tried before.

CHAPTER SIX

SEARCHING FOR OTHER EARTHS

There are only eight discovered planets in our solar system. But every single star in the night sky is a potential sun. In fact, scientists have found evidence that planets located outside our solar system, called exoplanets, orbit most stars. For a long time, the idea of exoplanets was purely theoretical. Scientists believed they were out there, but no one had ever found one. That all changed in the 1990s, when several groups of scientists detected distant worlds for the first time. In response to the discoveries, *New York Times* journalist John Noble Wilford wrote in 1997, "Other worlds are no longer the stuff of dreams and philosophic musings. They are out there, beckoning, with the potential to change forever humanity's perspective on

Artists have created concept images of what exoplanets could look like, based on observations of their diameters, masses, and distances from their stars.

its place in the universe."[1] Some space scientists have made it their life's work to search for and study exoplanets.

FLOATING FREE

Not all planets circle a star. Astronomers have discovered orphan planets, also called rogue planets, that float through space independent of a star. In 2013, astronomers took a picture of one. "It has all the characteristics of young planets found around other stars, but it is drifting out there all alone," says Michael Liu of the Institute for Astronomy at the University of Hawaii. "I had often wondered if such solitary objects exist, and now we know they do."[2] In 2017, a group of astronomers estimated that there is one Jupiter-sized orphan planet for every four stars in the galaxy.

Scientists estimate the Milky Way galaxy is home to more than 100 billion planets.

DIPS, WOBBLES, AND CURVES

Astronomers see the planets in this solar system easily through telescopes or even with the naked eye. But they can't see most exoplanets this way. These worlds are much too distant, and the stars they orbit are much too bright. Finding an exoplanet is like trying to make out a moth that's flying around a bright flood lamp many miles away. Even with advanced telescopes, that's not always possible. So space scientists look for indirect evidence of exoplanets. Most of the time, they look for patterns of change in the brightness or motion of a star.

A planet that circles a star will regularly block some of the star's light as it passes in front of the star from the viewer's perspective. If the orbits are lined up the right way as viewed from Earth, astronomers will see regular dips in brightness, which are called transits. A planet's gravity will also make a star wobble a tiny bit

The passage of an exoplanet in front of its star can reveal the planet's existence to human observers.

Like all spacecraft, the Kepler telescope underwent extensive testing and checks on Earth before being launched into space.

back and forth, and astronomers can watch for these wobbles. A third technique looks at how light from a distant star travels past a star that's closer to Earth. The gravity of the near star and any planets orbiting it will make the distant starlight bend slightly. The nature of this bend reveals whether exoplanets are getting in the way. Scientists can measure all these dips, wobbles, and curves to figure out information about an exoplanet, often including its size and distance from its star.

ON THE HUNT

By the 2000s, astronomers had found dozens of exoplanets. Most of them were gas giants somewhat similar to Jupiter or Saturn. Then, in 2009, exoplanet hunting shifted into high gear with the launch of the Kepler Space Telescope. It observed hundreds of thousands of stars, watching for transits. After Kepler's mission ended in 2018, a new telescope took over. The Transiting Exoplanet Survey Satellite (TESS) has followed in Kepler's footsteps, searching for transits across an even vaster region of the sky than Kepler could see.

Thanks mainly to data from these telescopes, the number of known exoplanets has risen sharply, to around 4,000 and counting.[3] "Kepler has surprised pretty much every single person alive—at least everyone working on exoplanets," says Sara Seager, an astrophysicist at the Massachusetts Institute of Technology (MIT).[4] Human astronomers haven't been able to keep up with all the data Kepler and TESS have gathered. Computer software helps churn through it all, flagging transits. Researchers have also organized citizen science groups, including Planet Hunters and Exoplanet Explorers, to allow ordinary people to help sort through the data. "People anywhere can log on and learn what real signals from exoplanets look like, and then look through actual data collected from the Kepler telescope to vote on whether or not to

AIR TO BREATHE

Some space scientists are looking for signs of life in distant planets' atmospheres. On Earth, the rise of life had a huge impact on the atmosphere. Algae and other plants filled the air with oxygen. Other life forms, including humans, evolved to breathe this oxygen. Finding oxygen in an exoplanet's atmosphere might indicate the presence of life. But scientists have found other nonliving processes that can also produce oxygen. They're looking for more subtle ways to tell a living planet's atmosphere from a nonliving one.

"Hundreds or thousands of years from now, when people look back at our generation, they will remember us for being the first people who found the Earth-like worlds."[5]

—*Sara Seager, astronomer, Massachusetts Institute of Technology*

classify a given signal as a transit or just noise," says Jessie Christiansen, an astrophysicist at the California Institute of Technology.[6]

THE GOLDILOCKS ZONE

Every new exoplanet is an exciting discovery. But most planet hunters hope to find worlds like Earth. These have the most potential for possible future colonization, and they may even already be home to some form of alien life. "My life's obsession for planets is to find another planet like Earth, one with water and continents and with signs of life in the atmosphere," says Seager.[7] A potentially habitable Earthlike planet should be similar in size and composition to Earth. It should have an atmosphere and liquid water. For liquid water to exist, the planet should orbit at a certain distance from its sun, in a region astronomers have nicknamed the Goldilocks zone. If the planet orbits too close, any water on

Artists have speculated on what the surfaces of distant exoplanets might look like.

its surface would evaporate. It also can't orbit too far away, or the water would all freeze. Surface temperatures can't be too hot or too cold—they must be just right.

In 2014, researchers announced the discovery of the first Earth-sized exoplanet orbiting in the Goldilocks zone of its star. The planet is called Kepler 186f. The star it orbits is in the constellation Cygnus and is much smaller and dimmer than the sun. So while the planet "is not an Earth twin, it is perhaps an Earth cousin," says Tom Barclay of the Bay Area Environmental Research Institute, who led the team that found the planet in Kepler's data.[8] Many other Earthlike exoplanets have been discovered since. In 2016, the group Planet Explorers made a discovery in Kepler's data. It noticed that a star Kepler observed seemed to have a planet orbiting around it. Scientists took a closer look at the data and confirmed the find. "We, the science team, initially missed this signal," says lead researcher Adina Feinstein of the University of Chicago. "It took the eyes and excitement of the citizen scientists to draw our attention to this new planet."[9] The planet, K2-288Bb, is about twice the size of Earth

BIGGER, BETTER TELESCOPES

Astronomers who study exoplanets are excited about a new generation of telescopes. These will help provide better images and data than ever before on distant worlds.
- James Webb Space Telescope: Expected to begin operating in 2021. This space telescope will be able to collect data about the atmospheres of some exoplanets.
- Large Synoptic Survey Telescope in Chile: Expected to begin operating in 2023. This ground telescope will collect frequent images of the entire sky over several years, allowing astronomers to detect small changes over time.
- Extremely Large Telescope in Chile: Expected to begin operating in 2024. This ground telescope should be able to take pictures of exoplanets similar to Earth in size.
- Giant Magellan Telescope in Chile: Expected to begin operating in 2025. The images this ground telescope collects will be ten times sharper than those of the Hubble Space Telescope.

and orbits in the Goldilocks zone. However, scientists don't know if either of these planets has an atmosphere or what their surfaces look like.

LOOKING FOR ALIEN LIFE

To really understand whether an exoplanet might be a place that humans, bacteria, or other forms of life could call home, scientists need to study its composition and its atmosphere. The best way to do that is to take direct pictures of the planet. That's not always possible. But it has been done. In 2008, astronomer Christian Marois of the Herzberg Institute of Astrophysics in Canada and his team announced the first images ever taken of a system of exoplanets. The team used a computer image processing technique that allowed them to remove most of the light from the bright star in order to see the planets

In 2018, Sara Seager discussed an upcoming exoplanet-hunting mission.

A direct image of the exoplanet 2M1207b, *in red*, orbiting its star approximately 230 light years from Earth

around it. Marois was on an airplane going through observations he'd taken with two different telescopes when he made the discovery. "I just wanted to get up and yell in the plane, 'Yes! We found something!'" he said. At first, they only saw three planets. But then in late 2010, they found a fourth. All four of them are huge—larger than Jupiter. "It's been a planet adventure," says Marois.[10]

Another team of scientists studied the atmosphere of one of those four giant planets. They found blistering hot temperatures and a global storm of iron and dust. "Our observations suggest a ball of gas illuminated from the interior, with rays of warm light swirling through stormy patches of dark clouds," says astronomer Sylvestre Lacour of the Paris Observatory in France.[11] This planet is not a place life could survive. But space scientists are continually coming up with new and improved ways to see distant planets more clearly. They are also building bigger and better telescopes. New techniques and equipment should make it possible to study the atmospheres of small, rocky planets, says Seager.

Perhaps someday, one of these efforts will find the first evidence of an Earthlike planet with liquid water or even life on its surface. Seager is one of many space scientists who are hoping to make the dream of finding life elsewhere in the universe a reality. "I think I do it because I was a born explorer," she says. "I start a project and I get really excited about it, the heart beats faster. I just love what I do."[12] Spacecraft may not be able to visit exoplanets yet, but space scientists are hard at work exploring them from afar.

The absorption spectra for oxygen

SCIENCE CONNECTION
A REVEALING RAINBOW

Scientists can use a technique called spectroscopy to turn the light from a planet into information about its atmosphere. "Spectroscopy I believe is the most fascinating part of modern astronomy because it enables you to tell something about the nature of the objects in the universe in much detail without going there and touching them," says Rolf-Peter Kudritzki, an astronomer at the University of Hawaii.[13]

All light is made up of waves of electromagnetic radiation. These waves come in many wavelengths. In visible light, each wavelength is a different color. The first step of spectroscopy is to split a single beam of light into a spectrum of all of the wavelengths it contains. This type of splitting is sometimes visible on Earth. When sunlight travels through a foggy sky or a crystal prism, it splits to form a rainbow.

Astronomers use a tool called a spectrograph to split the light from distant planets. However, they don't see a smooth rainbow of perfect colors. Gaps or dark lines appear in the spectrum. These gaps appear because some of the light is absorbed on its way past the planet. Elements and molecules—including oxygen, water vapor, and carbon dioxide—each absorb light in a unique way, marking a pattern of lines into the spectrum. This pattern is like a fingerprint identifying an element or molecule. Astronomers can use these fingerprints to figure out the elements and gases that must be present in an exoplanet's atmosphere.

CHAPTER SEVEN
THE EXPANDING UNIVERSE

Scientists send spacecraft to explore the solar system. They also use telescopes to peer at planets, moons, stars, exoplanets, and other objects in the distant reaches of the galaxy and beyond. They study everything from giant clouds of dust and gas called nebulae to exploding stars called supernovae. They've also found quasars, the brightest objects in the universe, and black holes, which are stars that became so large they collapsed in on themselves. Black holes have such intense gravity that not even light escapes. Scientists have evidence that every galaxy has a black hole at its center. Galaxies pack together into groups, clusters, and superclusters.

When scientists look at distant objects in space, they are looking into the distant past.

QUASARS

While a black hole is actively consuming material, that material speeds up and shines with extremely bright light as it falls toward the black hole. Jets of plasma also shoot from the sides of an active black hole. All this light forms a quasar, the brightest and most energetic type of object in the universe. Quasars produce new stars and likely even whole galaxies.

When astronomers peer far into distant space, they are also looking back in time. That's because light takes time to travel across space. The light that reaches Earth from very distant objects reveals events that actually happened millions or billions of years ago. Scientists called cosmologists study the most distant reaches of space in an attempt to understand the history and structure of the entire universe.

EVERYTHING FLIES APART

In 1929, astronomer Edwin Hubble made one of the most astonishing discoveries in space science. Every galaxy he observed was moving away from Earth. He realized that this meant the universe was getting larger. Space itself was stretching and expanding. However, for most of the 1900s, physicists believed that space would eventually stop getting bigger and might even shrink in the distant future. "Everybody had assumed that the universe would slow down in that expansion because gravity would attract everything to everything else," says Saul Perlmutter of Lawrence Berkeley National Laboratory.[1] Just as a ball thrown into the air eventually slows and then falls because of gravity, the universe seemed like it should slow down, too.

In the 1990s, Perlmutter's team and a separate team were each working to measure the rate of the universe's expansion. The two teams' data showed something very surprising.

The expansion was not slowing down over time—it was speeding up. Perlmutter and the two leaders of the other group shared the 2011 Nobel Prize in Physics for their discovery. "What we discovered was a huge surprise," says Perlmutter. "We have been comparing it to throwing an apple up in the air, and finding that it doesn't fall back to earth, but instead blasts off into outer space, mysteriously moving faster and faster."[2] To explain the accelerating expansion, physicists imagined a new force that they call dark energy. This energy supposedly works against gravity to push things apart.

No one has directly detected dark energy yet or really understands what causes it. Different groups have measured different rates for the expansion of the universe. And some have suggested that dark energy may not actually exist. They have pointed out that the structure of the universe is not smooth. So the rate of expansion might vary in different parts of the universe. The universe could be even more complex and amazing than anyone ever imagined.

EINSTEIN'S MISTAKE

Albert Einstein's theory of general relativity suggested that the universe should either get bigger or smaller. But the scientist thought the universe shouldn't be like that. It should stay still. So Einstein imagined a new force that he called the cosmological constant. A certain value for this force made a stationary universe possible. When Hubble discovered that the universe was, in fact, expanding, Einstein said he had made a big mistake in adding that extra force to his equations. But today, an additional force is back with a new name—dark energy. This time, though, scientists introduced the new force to help explain their observations.

BACK TO THE BIG BANG

An expanding universe must have been smaller in the past. Trace the history of the universe back far enough, and at the beginning, everything was most likely compressed

Powerful space telescopes have peered into the universe's first billion years.

into a single point. What happened next is known as the Big Bang. It's how most modern physicists believe the universe began. Somehow, a single point began expanding, and like a balloon blowing up, space has been getting bigger ever since.

However, even though telescopes allow scientists to see into the distant past, they cannot look as far back as the Big Bang. In the first years after the universe began around 13.8 billion years ago, light couldn't yet move. Everything was too hot and under too much pressure. At around 380,000 years after the Big Bang, light finally began to spread through the expanding universe.[3] This light, an echo of the Big Bang, has been traveling in all directions ever since, getting fainter and more and more stretched out as space expands. Yet scientists managed to detect the remnants of the universe's first light rays in 1964. They called the discovery the cosmic microwave background radiation. It fills all of outer space with a faint glow and provides evidence that the Big Bang happened.

> "Because the expansion is speeding up . . . astronomers in the far future, looking out into deep space, will see nothing but an endless stretch of static, inky, black stillness."[4]
> —Brian Greene, physicist, Columbia University

RIPPLES IN SPACETIME

The cosmic microwave background radiation marks the edge of the universe that scientists can observe. But a new technique may one day allow them to look even farther. September 14, 2015, marked the beginning of a new era in space science. This was the date that scientists detected gravitational waves for the first time. Just as waves spread across

> "With gravitational waves, we should be able to see all the way back to the beginning. . . . I'm positive that there are things out there that we've never seen that we may never be able to see and that we haven't even imagined—things that we'll only discover by listening."[5]
> —*Allan Adams, physicist, Massachusetts Institute of Technology*

HOW TO CATCH GRAVITATIONAL WAVES

LIGO is housed in two separate L-shaped buildings, one in Louisiana and one in Washington state. Each arm of the *L* protects a laser that is 2.5 miles (4 km) long. Each laser's job is to keep track of the position of a mirror to an extremely precise degree. When a gravitational wave passes through Earth, one laser beam gets slightly shorter and the other slightly longer. The difference in length is smaller than the width of an atom's nucleus. When this happens at both buildings at once, then scientists know they found a real signal.

the surface of a pond after something disturbs the surface, ripples can also spread through spacetime when massive objects collide, explode, or collapse. These gravitational waves stretch spacetime in one direction while squashing it in the other.

Famed physicist Albert Einstein's theory of general relativity predicted a century ago that waves like this should exist. But detecting them wasn't easy. By the time one of these waves gets to Earth, the amount of stretching and squashing that happens is extremely tiny. The waves pass through people and objects all the time without anyone noticing. The detector that scientists built, called the Laser Interferometer Gravitational-Wave Observatory (LIGO), consists of two sites. One is in Washington state, and the other is in Louisiana. LIGO found no signals at all during its first eight years of operations. Then the detector shut down in 2010 for upgrades. Engineers made the detector even more sensitive. Just a few days after they turned it on again in 2015, the team found the first signal. It was the echo of a collision of two black holes.

The black holes had swirled around each other and then collided. "The shock would have released more energy than the light from all the stars in the universe for that brief instant," says B. S. Sathyaprakash of Cardiff University.[6]

A NEW WAY OF LISTENING

Astronomers had never before detected an event like this. "We have observed the universe through light so far," says Alberto Vecchio, a LIGO researcher at the University of Birmingham in the United Kingdom. But light only reveals some of what happens in the cosmos. Gravitational waves allow scientists to witness completely different phenomena, he says. "We have opened a new way of listening . . . which will allow us to discover phenomena we have never seen before."[7] Plans are underway to build and launch a space-based gravitational wave detector called the Laser Interferometer Space Antenna (LISA) by the 2030s. It will be able to detect tinier changes than LIGO can.

Gravitational wave measurements may one day reveal the moments directly after the Big Bang. "At some point, not with the detectors we have now, we hope to be able to look at the beginnings of the universe," said Rainer Weiss of MIT, who shared the

PORTRAIT OF A BLACK HOLE

Black holes swallow everything nearby, even light. But before getting sucked in, light and other materials swirl around a black hole, forming a disk. Though the heart of a black hole is invisible, this disk glows very faintly. In 2019, scientists revealed a picture of the disk around a black hole for the first time. To create the picture, eight telescopes in locations all around the globe worked together, effectively acting as one huge, planet-sized telescope. Scientists had already proven that black holes existed. But the photograph was still an important breakthrough. As astrophysicist Avi Loeb of Harvard University says, "Seeing is believing."[8]

2017 Nobel Prize in Physics for the gravitational wave discovery.[9] It would not be possible to directly detect gravitational waves from the birth of the universe, because those waves would be as vast as the universe. But scientists believe they will be able to study the cosmic microwave background radiation to find very early gravitational echoes of the Big Bang.

Cosmologists plan to use the LIGO detector to settle the question of how quickly the universe is expanding. This will help to reveal how the universe might end. It could be a Big Rip, in which clusters of galaxies and solar systems get torn apart by ever-increasing expansion. Or the expansion could slow down, ending with a Big Chill, as galaxies drift ever farther apart and cool over time. Space scientists are searching for the answers to these very big questions.

The LIGO detectors are enormous scientific instruments.

The expansion of the universe has left telltale signs that astronomers have detected and deciphered.

SCIENCE CONNECTION
REDSHIFT

Saul Perlmutter and the teams that determined that the expansion of the universe was accelerating searched the sky for a certain type of supernova. This type of supernova always explodes in the same way and shines with the same brightness. Astronomers call this type of object a standard candle. Astronomers can measure how much the light dims to tell how far it has traveled.

But astronomers needed more than just distance to figure out how quickly the universe was expanding. They also needed to know how quickly these supernovae were moving away from Earth. They did this by measuring something called redshift.

When visible light waves get squished on their way to an observer, they turn bluer. When they get stretched while moving away, they turn redder. Since the universe is expanding, every part of it is moving away from every other part. So all of the stars and galaxies that astronomers see appear redder than normal. The amount of redshift tells astronomers how quickly distant objects are moving away.

"When the supernova explodes, it sends out mostly blue light," Perlmutter explains. "That blue light means a short wave length of light. The more it stretches, the more it starts to turn red. And that tells us the amount the universe stretched between the time of the explosion, and today."[10] Supernovae closer to Earth move away more quickly than more distant ones, suggesting that the universe's expansion is speeding up.

CHAPTER EIGHT
INTO THE UNKNOWN

Space scientists have learned many incredible things about our solar system, our galaxy, and the entire universe. But more great discoveries and expeditions in the field are still to come. In the near future, space scientists may find life on another planet or moon. They may send human explorers to Mars and robotic explorers to other star systems. Someday, in the much more distant future, humans may even set up colonies in space or on other planets or moons.

Some experts say that moving to outer space is the best way for humanity to survive into the distant future. That way, if some disaster happens on Earth, human civilization will not go extinct. "If humanity is to continue for another million years it relies on boldly going where no one has gone before," said the famous physicist Stephen Hawking in 2017.[1] In a separate speech that same year, he said, "We are running out of space

A human settlement on Mars is one long-term goal of many space scientists.

on Earth and we need to break through the technical limitations preventing us living elsewhere in the universe."[2]

The technical problems holding back human space exploration and colonization are not easy to solve. But a second space race has broken out in recent years as private companies including Arianespace, SpaceX, Blue Origin, and Virgin Galactic compete with each other to build improved rockets, spacecraft, and other spacefaring technology. The governments of China, India, Israel, Saudi Arabia, and United Arab Emirates are also investing in space exploration. China successfully landed a robotic rover on the moon in 2013, and it wants to build a moon base in the 2030s.

STEPHEN HAWKING

British physicist Stephen Hawking spent much of his career studying the properties of black holes. He also wrote popular science books to help communicate space science to the public, including the 1988 best-seller *A Brief History of Time*. Hawking made crucial contributions to cosmology, but he is also well known for his personal story. In the 1960s, he was diagnosed with a degenerative muscular disease. In later years he used a wheelchair and lost his ability to speak, but he continued to work and write. Hawking died in 2018.

DESTINATION MARS

"There's going to be a civilization on Mars. It's going to happen a lot sooner than people think," says Stephen Petranek, author of *How We'll Live on Mars*.[3] The problems this Mars civilization would have to solve are numerous. The planet lacks breathable air. The thin atmosphere lets through lots of harmful radiation from the sun and outer space. Mars is also incredibly cold. But Petranek and some other experts believe that it will be possible to transform Mars into a more Earthlike place through a process called terraforming. Thickening the atmosphere

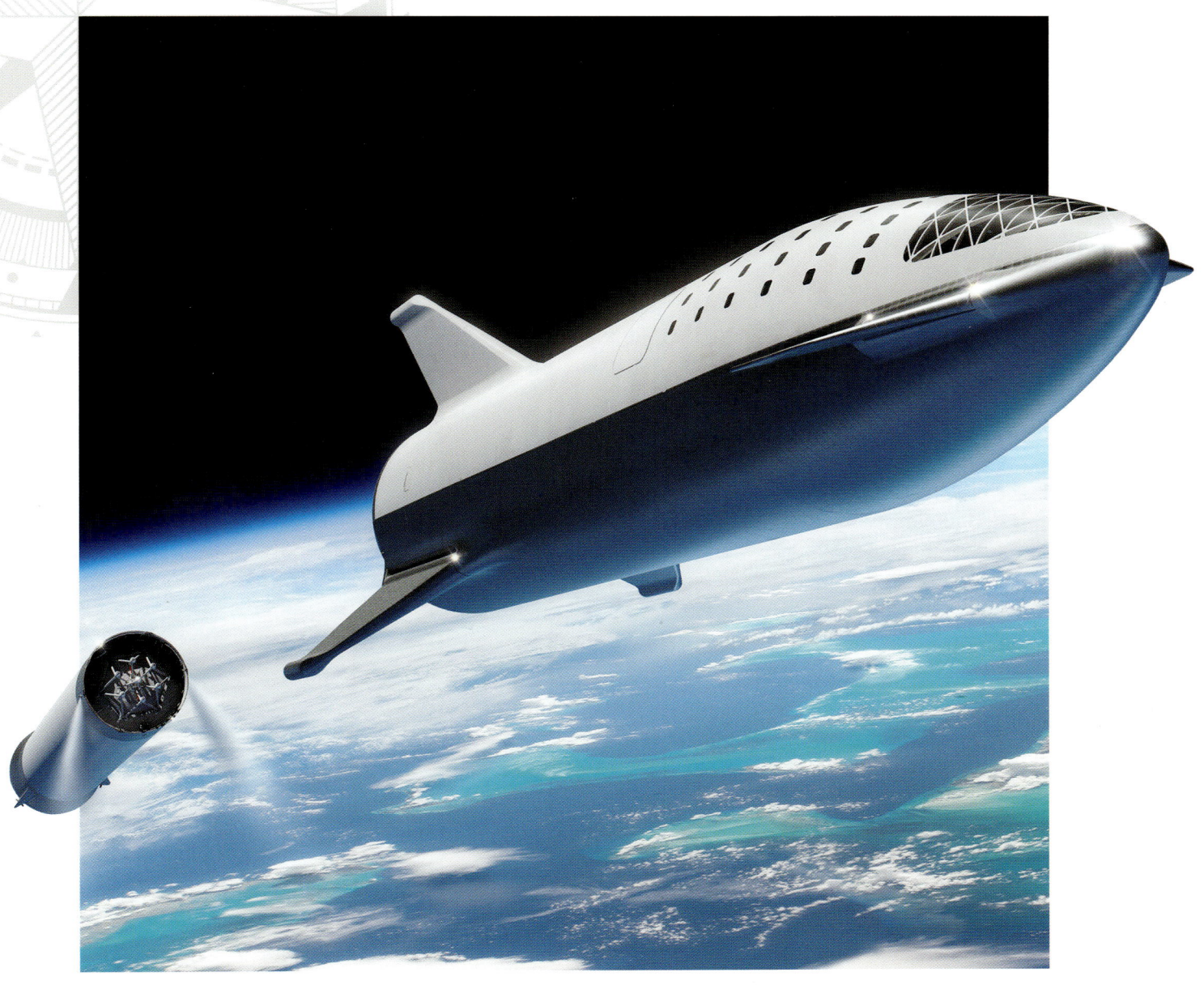

SpaceX is developing innovative reusable rockets and spacecraft that it hopes will drive down the cost of space travel.

Elon Musk has spoken frequently about his desire to land people on Mars.

could warm the planet and make conditions more comfortable for settlers. However, such a plan would take many years to accomplish and will likely only be possible in the distant future, if ever.

For now, a colony on Mars would have to make all of its own air, fuel, food, and heat. The company SpaceX is planning a series of missions to Mars starting in the 2020s. "This is not about everyone moving to Mars," says Elon Musk, founder of SpaceX. He says it's about reducing the risk of humans going extinct and "having a tremendous sense of adventure."[4]

On the first trip, two uncrewed cargo ships would bring the resources and machines needed for life support, a power supply, and mining systems. Space travel using rocket technology requires a lot of fuel. The mining systems would gather elements from the surface and convert them to fuel, allowing a spacecraft to refuel on the planet for a trip back to Earth. Mars-bound ships would also have to refuel after reaching Earth orbit. SpaceX is also planning to build and launch tanker ships that will carry only

SPACE VACATIONS

Companies including SpaceX, Blue Origin, and Virgin Galactic plan to fly tourists into space. Around 700 hopeful travelers have already paid $250,000 for a seat on one of the first Virgin Galactic space flights.[5] These will travel briefly into space, giving tourists a few minutes of weightlessness before returning to Earth. In 2023, SpaceX plans to carry Yusaku Maezawa, a billionaire from Japan, around the moon. Maezawa has not revealed how much he paid for the trip, but he plans to select six to eight artists to travel along with him for free. "They will be asked to create something after they return to Earth," said Maezawa. "These masterpieces will inspire the dreamer within all of us."[6]

"We're going to be a spacefaring civilization."[7]

—*Elon Musk, founder of SpaceX*

fuel. Once the cargo ships have brought enough supplies to Mars, humans would take the trip. While these first Mars explorers worked on setting up camp, they'd likely live on the spacecraft. Their lives wouldn't be easy. For a long time, they would have to rely on shipments from Earth for food, medical supplies, machinery, and everything else they couldn't yet build or grow on Mars. But they would make an incredible mark on history. They would usher humanity into a new era as a multi-planet species.

INTERSTELLAR TRAVEL

Most see Mars or the moon as the best location for an early human base or colony because both locations are relatively nearby. It takes around nine months to get to Mars with current technology. But planets or moons in other solar systems might prove even more hospitable than these. The planet Proxima b, located in the closest star system to Earth, Alpha Centauri, has the right size and distance from its star for life to survive. However, scientists have no idea if the planet has an atmosphere or liquid water. Even if it does, people have no workable way to get there. It takes four years for a beam of light to travel between Alpha Centauri and Earth. And according to the laws of physics, nothing can move faster

---

ELON MUSK

Elon Musk is a brilliant inventor and an ambitious dreamer. Born and raised in South Africa, he made a fortune as a young man after founding and selling the company PayPal. He decided to invest his time and money into two futuristic industries, electric cars and commercial rockets. He runs electric car company Tesla Motors and rocket company SpaceX. In 2018, SpaceX successfully launched its Falcon Heavy rocket for the first time. Test flights like this usually have a payload of worthless weights to simulate the mass of a spacecraft. But Musk instead placed his Tesla convertible atop the rocket. The launch succeeded, and the red sports car is now orbiting the sun.

than light. A modern spacecraft, which can't travel anywhere near light speed, would take around 6,300 years to reach Proxima b.[8]

But space scientists are hard at work on new technology that may make faster travel possible, if not for humans, then at least for tiny robots. The Breakthrough Starshot Initiative is planning for a mission to send a fleet of 1,000 miniature probes called StarChips all the way to Proxima b using a method called light sailing. The tiny probes would use light, thin sails to catch a powerful beam of laser light. The laser would accelerate the tiny craft to the incredible speed of 134 million miles per hour (216 million km/h).[9] That's a thousand times faster than any human-built object has traveled. At that speed, the probes could reach Proxima b in around 20 years. Yuri Milner, a technology entrepreneur and billionaire from Russia, is backing the project. At a 2016 announcement about the initiative, Stephen Hawking said, "The limit that confronts us now is the great void between us and the stars. But now we can transcend it. . . . Today we commit to this next great leap into the cosmos, because we are human and our nature is to fly."[10]

SHOULD PEOPLE LEAVE MARS ALONE?

Some experts say that the idea of using Mars as a backup planet for humanity is a big mistake. Trying to survive on Mars would be much more expensive and difficult than fixing any crisis on Earth. "For anyone to tell you that Mars will be there to back up humanity is like the captain of the *Titanic* telling you that the real party is happening later on the lifeboats," says Lucianne Walkowicz, an astronomer at Adler Planetarium in Chicago.[11] Settlers would also permanently erase many of Mars's beautiful features to replace them with human habitats. And they'd contaminate the Martian environment with trash and microbes. This could damage anything that may already live there. "If there is life on Mars," wrote Carl Sagan in his book *Cosmos*, "Mars then belongs to the Martians, even if the Martians are only microbes."[12]

A fleet of microchip-sized probes could someday be humanity's first ambassador to another star.

Space has been called the final frontier. The desire to explore this frontier pushes space scientists to dream, invent, and engineer their way toward exciting discoveries. With the help of robotic rovers, space probes, telescopes, detectors, and many other tools, space scientists are expanding human knowledge. It's impossible to know what secrets the universe still keeps from humanity, or how far space exploration may someday reach. But endless possibilities await out among the stars.

"Spreading out into space will completely change the future of humanity. It may also determine whether we have any future at all."[13]

—Stephen Hawking, physicist, University of Cambridge

ESSENTIAL FACTS

SIGNIFICANT EVENTS

- Edwin Hubble discovered that the universe was expanding in 1929. In the 1990s, scientists discovered that the expansion of the universe is speeding up. They named the mysterious force responsible for the expansion *dark energy*.

- On April 12, 1961, Yuri Gagarin became the first human being to orbit Earth. On July 20, 1969, Neil Armstrong and Buzz Aldrin became the first people to walk on the moon.

- In the 1990s, several teams of scientists detected planets in other solar systems for the first time. Since then, the number of known exoplanets has risen into the thousands.

- On September 14, 2015, scientists detected ripples in spacetime, called gravitational waves, for the first time. The waves came from the collision of two black holes.

KEY PLAYERS

- *Curiosity* is a robotic rover that landed on Mars in 2012. The rover has made many important discoveries, including the confirmation that water once flowed on the planet.

- Scott Kelly is an American astronaut who spent a year in space to help study the effects of the outer space environment and microgravity on the human body.

- Elon Musk is an inventor, entrepreneur, and founder of the company SpaceX. He has made it his mission to establish a human presence on Mars.

IMPACT ON SCIENCE

Space scientists explore the cosmos to better understand big questions, including how the universe came to exist and what else is out there beyond Earth. Especially exciting areas of current research include the search for signs of life in this solar system and for potentially habitable worlds in other star systems using telescopes, probes, and rovers. Gravitational wave detectors are making it possible to study black holes and other phenomena, including the Big Bang and the expansion of the universe, in new ways.

QUOTE

"I feel like a little kid, like a sorcerer, like the luckiest person alive. I am in space, weightless, and getting here only took 8 minutes and 42 seconds. Give or take a few thousand days of training."

—Chris Hadfield, astronaut

GLOSSARY

asteroid
A rocky object that orbits the sun, typically within the asteroid belt between Mars and Jupiter.

atmosphere
The layer of gases surrounding a planet.

comet
An object that develops a tail of gas and dust as it swings past the sun on its orbit through the solar system.

cosmologist
A scientist who studies the nature, history, and future of the universe.

exoplanet
A planet located in another star system.

galaxy
A group of millions or billions of stars.

microbe
A microscopic living thing.

orbit
The curved path of a celestial object or spacecraft around a star, planet, or moon.

organic
Containing the element carbon.

radiation
A form of energy that can damage living tissue.

rover
A wheeled robotic vehicle that explores a planet or moon and returns data to Earth.

satellite
An object, often human-made, orbiting a planet or moon that is larger than itself.

spectrum
A series of colors or wavelengths of light separated from a single beam.

wavelength
The distance between two crests in a wave, such as one of light or sound. A change in wavelength alters the color of light or the pitch of sound.

ADDITIONAL RESOURCES

SELECTED BIBLIOGRAPHY

Melvin, Leland. "An Astronaut's Story of Curiosity, Perspective and Change." *TED*, Nov. 2018, ted.com. Accessed 21 Apr. 2019.

Peterson, Carolyn Collins. *Space Exploration: Past, Present, Future.* Amberley, 2018. Print.

Pyle, Rod. *Interplanetary Robots: True Stories of Space Exploration.* Prometheus, 2019. Print.

Raz, Guy. "Peering Deeper into Space." *NPR*, 9 Feb. 2018, npr.org. Accessed 25 July 2019.

Sample, Ian, and Hannah Devlin. "'A New Way to Study Our Universe': What Gravitational Waves Mean for Future Science." *Guardian*, 3 Oct. 2017, theguardian.com. Accessed 25 July 2019.

FURTHER READINGS

Kruesi, Liz. *Astronomy.* Abdo, 2016.

Kruesi, Liz. *Space Exploration.* Abdo, 2015.

Peake, Tim. *Ask an Astronaut: My Guide to Life in Space.* Little, Brown and Company, 2017.

Space!: The Universe as You've Never Seen It Before. DK Publishing, 2015.

ONLINE RESOURCES

To learn more about space scientists, please visit **abdobooklinks.com** or scan this QR code. These links are routinely monitored and updated to provide the most current information available.

MORE INFORMATION

For more information on this subject, contact or visit the following organizations:

NATIONAL AERONAUTICS AND SPACE ADMINISTRATION (NASA)

300 E Street SW, Suite 5R30

Washington, DC 20546

202-358-0001

nasa.gov

NASA runs the space program in the United States. This organization recruits and trains astronauts, manages missions to space, and develops new technology for space exploration.

THE PLANETARY SOCIETY

60 S. Los Robles Ave.

Pasadena, CA 91101

626-793-5100

planetary.org

The Planetary Society works to empower ordinary citizens to help advance space exploration and better understand the cosmos.

SOURCE NOTES

CHAPTER 1. EXPLORING MARS
1. Fraser Cain. "How Long Does It Take to Get to Mars?" *Universe Today*, 9 May 2013, universetoday.com. Accessed 14 Aug. 2019.
2. Mike Wall. "Touchdown! Huge NASA Rover Lands on Mars." *Space*, 6 Aug. 2012, space.com. Accessed 5 Apr. 2012.
3. Jesus Diaz. "The Mars Curiosity Rover Has LANDED!—Live Coverage from JPL." *Gizmodo*, 6 Aug. 2012, gizmodo.com. Accessed 5 Apr. 2019.
4. Ian Sample. "Adam Steltzner: 'I Could Not Imagine the Curiosity Landing Working.'" *Guardian*, 30 Dec. 2012, theguardian.com. Accessed 5 Apr. 2019.
5. "NASA's Opportunity Rover Mission on Mars Comes to End." *NASA Mars Exploration Program*, 13 Feb. 2019, mars.nasa.gov. Accessed 14 Aug. 2019.
6. Mike Wall. "Ancient Mars Lake Could Have Supported Life, Curiosity Rover Shows." *Space*, 9 Dec. 2013, space.com. Accessed 5 Apr. 2019.
7. Ryan P. Smith. "Think Mountain Time's Confusing? Try Living on Martian Time." *Smithsonian*, 11 Jan. 2018, smithsonianmag.com. Accessed 3 May 2019.
8. Nola Taylor Redd. "Water on Mars: Exploration & Evidence." *Space*, 18 Aug. 2018, space.com. Accessed 5 Apr. 2019.
9. "Astronaut/Cosmonaut Statistics." *World Space Flight*, 14 Aug. 2019, worldspaceflight.com. Accessed 14 Aug. 2019.
10. Nadia Drake. "They Saw Earth from Space. Here's How It Changed Them." *National Geographic*, Mar. 2018, nationalgeographic.com. Accessed 14 Aug. 2019.
11. "NASA Rover Finds Conditions Once Suited for Ancient Life on Mars." *NASA*, 12 Mar. 2013, nasa.gov. Accessed 14 Aug. 2019.
12. Sample, "Adam Steltzner."

CHAPTER 2. THE RACE TO SPACE
1. Michael J.I. Brown. "Copernicus' Revolution and Galileo's Vision." *Conversation*, 30 May 2016, theconversation.com. Accessed 21 Apr. 2019.
2. Fordyce Williams. "Frequently Asked Questions about Dr. Robert H. Goddard." *Clark University*, n.d., Accessed 21 Apr. 2019.
3. "Robert Goddard: A Man and His Rocket." *NASA*, 9 Mar. 2004, nasa.gov. Accessed 14 Aug. 2019.
4. "De Forest Says Space Travel Is Impossible." *Lewiston Morning Tribune*, 25 Feb. 1957, news.google.com. Google News. Accessed 21 Apr. 2019.
5. Nola Taylor Redd. "Yuri Gagarin: First Man in Space." *Space*, 12 Oct. 2018, space.com. Accessed 21 Apr. 2019.
6. "Soviet Traveller Returns from Out of This World." *Life*, 21 Apr. 1969. *Google Books*, books.google.com. Accessed 11 Apr. 2019.
7. John F. Kennedy. "Excerpt from the 'Special Message to the Congress on Urgent National Needs.'" *NASA*, 24 May 2004, nasa.gov. Accessed 21 Apr. 2019.
8. Amy Shira Teitel. "How NASA Broadcast Neil Armstrong Live from the Moon." *Popular Science*, 5 Feb. 2016, popsci.com. Accessed 21 Apr. 2019.
9. "1969 Moon Landing." *History*, 23 Aug. 2018, history.com. Accessed 21 Apr. 2019.
10. "1969 Moon Landing."
11. Elizabeth Howell. "Eugene Cernan: Last Man on the Moon." *Space*, 16 Jan. 2017, space.com. Accessed 14 Aug. 2019.

CHAPTER 3. LIFE AS AN ASTRONAUT
1. Chris Hadfield. "What I Learned from Going Blind in Space." *TED*, Mar. 2014, ted.com. Accessed 21 Apr. 2019.

2. Hadfield, "What I Learned from Going Blind in Space."

3. Leland Melvin. "An Astronaut's Story of Curiosity, Perspective and Change." *TED*, Nov. 2018, ted.com. Accessed 21 Apr. 2019.

4. Chris Hadfield. *An Astronaut's Guide to Life on Earth*. Little, Brown and Company, 2013. 26.

5. Johnny Dodd. "NFL Player Turned Astronaut Leland Melvin Shares His 'Love Affair' with Earth in Nat Geo Doc." *People*, 27 May 2018, people.com. Accessed 21 Apr. 2019.

6. Marina Koren. "Sunrise, Sunset, Sunrise, Sunset, Sunrise." *Atlantic*, 15 Sept. 2015, theatlantic.com. Accessed 14 Aug. 2019.

7. Samantha Mathewson. "How Recycled Astronaut Pee Boosts Chances for Future Deep-Space Travel." *Space*, 16 Nov. 2016, space.com. Accessed 14 Aug. 2019.

8. Mike Massimo. *Spaceman: An Astronaut's Unlikely Journey to Unlock the Secrets of the Universe*. Crown Archetype, 2016. 179.

9. Marina Koren. "What a Year in Space Did to Scott Kelly." *Atlantic*, 11 Apr. 2019, theatlantic.com. Accessed 14 Aug. 2019.

10. Nancy Atkinson. "Spacewalking: Through an Astronaut's Eyes." *Universe Today*, 16 Mar. 2010, universetoday.com. Accessed 14 Aug. 2019.

11. Lori Byrd-McDevitt. "Describing Space—The Astronaut's Ultimate Dilemma." *Children's Museum*, 27 July 2016, childrensmuseum.org. Accessed 14 Aug. 2019.

12. Lulu Garcia-Navarro. "'Every Day Is a Good Day When You're Floating': Anne McClain Talks Life in Space." *NPR*, 17 Feb. 2019, npr.org. Accessed 21 Apr. 2019.

13. Jennifer Wall. "What Is Microgravity?" *NASA*, 16 June 2015, nasa.gov. Accessed 14 Aug. 2019.

14. "How Fast Does the Space Station Travel?" *Cool Cosmos*, n.d., coolcosmos.ipac.caltech.edu. Accessed 14 Aug. 2019.

CHAPTER 4. MYSTERIOUS MOONS

1. Nadia Drake. "These Are Some of the Solar System's Biggest Surprises." *National Geographic*, 27 Mar. 2014, nationalgeographic.com. Accessed 14 Aug. 2019.

2. Jesse Emspak. "Does Jupiter's Moon Europa Have a Subsurface Ocean? Here's What We Know." *Space*, 26 Sept. 2016, space.com. Accessed 14 Aug. 2019.

3. "Hubble: Possible Water Plumes on Jupiter's Moon Europa." *NASA/JPL*, 26 Sept. 2016, jpl.nasa.gov. Accessed 14 Aug. 2019.

4. Carl Sagan. "A Pale Blue Dot." *Planetary Society*, 2019, planetary.org. Accessed 14 Aug. 2019.

5. Drake, "These Are Some of the Solar System's Biggest Surprises."

6. "A Dozen New Moons of Jupiter Discovered, Including One 'Oddball.'" *Carnegie Institution for Science*, 16 July 2018, carnegiescience.edu. Accessed 14 Aug. 2019.

7. Carolyn Porco. "Could a Saturn Moon Harbor Life?" *TED*, Feb. 2009, ted.com. Accessed 14 Aug. 2019.

8. Porco, "Could a Saturn Moon Harbor Life?"

9. Drake, "These Are Some of the Solar System's Biggest Surprises."

10. "The Discovery of Eris." *Astrobiology Magazine*, 1 Mar. 2007, astrobio.net. Accessed 14 Aug. 2019.

11. Kathryn Hulick. "Person to Discover: Linda Spilker." *Odyssey*, vol. 23 no. 7, Sept. 2014.

12. Paul Scott Anderson. "Planet 9 Hypothesis Gets a Boost." *EarthSky*, 3 Mar. 2019, earthsky.org. Accessed 14 Aug. 2019.

13. "How Does NASA Communicate with Spacecraft?" *NASA Space Place*, n.d., spaceplace.nasa.gov. Accessed 26 Apr. 2019.

SOURCE NOTES CONTINUED

CHAPTER 5. CATCHING COMETS AND ASTEROIDS

1. "Rosetta: The Human Story." *Open University*, n.d., open.ac.uk. Accessed 26 Apr. 2019.
2. Sarah Knapton. "Rosetta Comet Landing: As It Happened." *Telegraph*, 12 Nov. 2014, telegraph.co.uk. Accessed 14 Aug. 2019.
3. "Rosetta: The Human Story."
4. Sheyna E. Gifford. "Astro Update: Rosetta's Comet." *Astrobiology Magazine*, 18 Oct. 2016, astrobio.net. Accessed 14 Aug. 2019.
5. Knapton, "Rosetta Comet Landing."
6. "Rosetta Finale Set for 30 September." *European Space Agency*, 30 June 2016, esa.int. Accessed 14 Aug. 2019.
7. Elizabeth Howell. "Hayabusa: Troubled Sample-Return Mission." *Space*, 31 Mar. 2018, space.com. Accessed 14 Aug. 2019.
8. Guy Raz. "Phil Plait: How Can We Defend Earth from Asteroids?" *TED Radio Hour, NPR*, 15 Feb. 2013, npr.org. Accessed 14 Aug. 2019.
9. Chandler Thornton and Euan McKirdy. "Japan Successfully Lands Robot Rovers on an Asteroid's Surface." *CNN*, 23 Sept. 2018, cnn.com. Accessed 14 Aug. 2019.

CHAPTER 6. SEARCHING FOR OTHER EARTHS

1. John Noble Wilford. "In a Golden Age of Discovery, Faraway Worlds Beckon." *New York Times*, 9 Feb. 1997, nytimes.com. Accessed 14 Aug. 2019.
2. Alan Boyle. "Astronomers Say They've Spotted Lonesome Planet without a Sun." *NBC News*, 9 Oct. 2013, nbcnews.com. Accessed 2 May 2019.
3. "How Many Exoplanets Are There?" *Exoplanet Exploration*, 2 May 2019, exoplanets.nasa.gov. Accessed 2 May 2019.
4. Sara Seager. Personal Interview. 7 July 2011.
5. Corey S. Powell. "Sara Seager's Tenacious Drive to Discover Another Earth." *Smithsonian*, May 2014, smithsonianmag.com. Accessed 2 May 2019.
6. Calla Cofield. "Five-Exoplanet System Discovered Thanks to Citizen Scientists." *Space*, 14 Jan. 2018, space.com. Accessed 14 Aug. 2019.
7. "Sara Seager: How Close Are We to Finding Life on Another Planet?" *WOSU*, 9 Feb. 2018, radio.wosu.org. Accessed 14 Aug. 2019.
8. Miriam Kramer. "5 Things to Know about Alien Planet Kepler-186f, 'Earth's Cousin.'" *Space*, 18 Apr. 2014, space.com. Accessed 14 Aug. 2019.
9. Corey S. Powell. "Citizen Scientists Discover Strange New World That Pro Astronomers Missed." *NBC News*, 10 Jan. 2019, nbcnews.com. Accessed 14 Aug. 2019.
10. Christian Marois. Personal Interview. 8 July 2011.
11. Michelle Starr. "We Just Got the Most Detailed Direct Observation of an Exoplanet Yet, and It's Brutal." *ScienceAlert*, 28 Mar. 2019, sciencealert.com. Accessed 2 May 2019.
12. Powell, "Sara Seager's Tenacious Drive."
13. "Astronomy—Spectroscopy—1/3." *YouTube*, uploaded by rhcrcgvp, 11 July 2009, youtube.com. Accessed 14 Aug. 2019.

CHAPTER 7. THE EXPANDING UNIVERSE

1. "Discovering the Accelerating Universe." *PBS LearningMedia*, 2015, mass.pbslearningmedia.org. Accessed 3 May 2019.
2. Marlene Cimons. "Scientist Who Helped Discover the Expansion of the Universe Is Accelerating." *National Science Foundation*, 3 Feb. 2015, nsf.gov. Accessed 3 May 2019.

3. Fraser Cain. "Why Can't We See the Big Bang?" *Universe Today*, 23 Oct. 2014, universetoday.com. Accessed 14 Aug. 2019.

4. Guy Raz. "Brian Greene: How Did a Mistake Unlock One of Space's Mysteries?" *TED Radio Hour, NPR*, 15 Feb. 2013, npr.org. Accessed 3 May 2019.

5. Allan Adams. "What the Discovery of Gravitational Waves Means." *TED*, Feb. 2016, ted.com. Accessed 3 May 2019.

6. Tim Radford. "Gravitational Waves: Breakthrough Discovery after a Century of Expectation." *Guardian*, 11 Feb. 2016, theguardian.com. Accessed 14 Aug. 2019.

7. Radford, "Gravitational Waves."

8. Maria Temming. "How Scientists Took the First Picture of a Black Hole." *Science News*, 11 Apr. 2019, sciencenews.org. Accessed 14 Aug. 2019.

9. Ian Sample and Hannah Devlin. "'A New Way to Study Our Universe': What Gravitational Waves Mean for Future Science." *Guardian*, 3 Oct. 2017, theguardian.com. Accessed 14 Aug. 2019.

10. Cimons, "Scientist Who Helped Discover the Expansion of the Universe Is Accelerating."

CHAPTER 8. INTO THE UNKNOWN

1. Sarah Knapton. "Human Race Is Doomed If We Do Not Colonise the Moon and Mars, Says Stephen Hawking." *Telegraph*, 20 June 2017, telegraph.co.uk. Accessed 14 Aug. 2019.

2. John Von Radowitz. "Stephen Hawking Has a Chilling Message about the Survival of Humanity." *Independent*, 20 May 2017, independent.co.uk. Accessed 14 Aug. 2019.

3. Guy Raz. "Stephen Petranek: How Will Humans Live On Mars?" *TED Radio Hour, NPR*, 21 Dec. 2018, npr.org. Accessed 3 May 2019.

4. Nadia Drake. "Elon Musk: A Million Humans Could Live on Mars by the 2060s." *National Geographic News*, 27 Sept. 2016, news.nationalgeographic.com. Accessed 14 Aug. 2019.

5. Jonathan O'Callaghan. "Blue Origin Launches Its Space Tourist Rocket for a 10th Time, with Plans to Fly Humans This Year." *Forbes*, 23 Jan. 2019, forbes.com. Accessed 3 May 2019.

6. Ciara Nugent. "Meet the Japanese Billionaire Who's Paying Elon Musk for a Trip to the Moon." *Time*, 18 Sept. 2018, time.com. Accessed 3 May 2019.

7. Drake, "Elon Musk."

8. "This Is How Many People We'd Have to Send to Proxima Centauri to Make Sure Someone Actually Arrives." *MIT Technology Review*, 22 June 2018, technologyreview.com. Accessed 3 May 2019.

9. Tim Radford. "Stephen Hawking and Yuri Milner Launch $100m Star Voyage." *Guardian*, 12 Apr. 2016, theguardian.com. Accessed 12 Apr. 2016.

10. Radford, "Stephen Hawking and Yuri Milner."

11. Guy Raz. "Lucianne Walkowicz: Should We Be Using Mars as a Backup Planet?" *TED Radio Hour, NPR*, 21 Dec. 2018, npr.org. Accessed 3 May 2019.

12. Zahaan Bharmal. "The Case against Mars Colonisation." *Guardian*, 28 Aug. 2018, theguardian.com. Accessed 14 Aug. 2019. August 28, 2018.

13. Knapton, "Human Race Is Doomed."

INDEX

ancient astronomers, 16
Apollo 11, 22
Armstrong, Neil, 22
asteroids, 18, 46, 52, 54, 58–63
astronauts, 20–22, 25, 26–38

Bennu, 60
Big Bang, 81–83, 85, 87
black holes, 13, 78, 80, 84, 85, 92
Blue Origin, 92, 95
Breakthrough Starship Initiative, 97
Brown, Michael, 47–48

Canadian Space Agency, 25
Cassini, 43–46, 50
Ceres, 59
Cernan, Eugene, 25
Challenger disaster, 25
Columbia disaster, 25
comets, 18, 47, 52–58, 60
Copernicus, Nicolaus, 16
Curiosity, 4–9, 12, 13, 50

Dawn, 59
de Forest, Lee, 20
Deep Space Network, 50
dwarf planets, 46–48, 52, 59

Einstein, Albert, 81, 84
Enceladus, 43–45
Eris, 47–48
Eros, 58
Europa, 42–43, 45
Europa Clipper, 45
European Space Agency (ESA), 12, 13, 25, 54, 58
ExoMars, 13
exoplanets, 64–75, 76, 78
expansion of the universe, 80–81, 83, 87, 88
Extremely Large Telescope, 72

Gagarin, Yuri, 21
Galilei, Galileo, 18
Galileo, 42
Giant Magellan Telescope, 72
Goddard, Robert, 19
Goldilocks Zone, 70–73
gravitational waves, 83–87

Hadfield, Chris, 26, 28, 36
Hawking, Stephen, 90–92, 97, 99
Hayabusa, 59, 60
Hayabusa2, 60
Hubble, Edwin, 80, 81
Hubble Space Telescope, 42, 72
human computers, 22
Huygens, 45–46

InSight, 13
International Space Station (ISS), 25, 31–37, 38
  construction, 25
  maintenance, 35–36
  research, 33–35
  resource recycling, 31
interstellar travel, 96–97

James Webb Space Telescope, 72
Japanese Aerospace Exploration Agency (JAXA), 25, 59–63
Jupiter, 18, 40, 42, 43, 54, 60, 66, 68, 75

Kelly, Scott, 32
Kennedy, John F., 21
Kepler, Johannes, 18
Kepler Space Telescope, 68–69, 72
Kuiper Belt, 47, 48, 50

Large Synoptic Survey Telescope, 72
Laser Interferometer Gravitational-Wave Observatory (LIGO), 84–87

Marois, Christian, 73–75
Mars, 4–13
   failed missions, 12
   Gale Crater, 8
   human missions, 92–96
   Mount Sharp, 8
   organic molecules, 12
   water, 9–11
Mars Express, 11
Mars 2020, 13
Marsh, Tom, 57
Massimo, Mike, 31–32
McClain, Anne, 36, 38
Melvin, Leland, 28
microgravity, 31, 33–35, 38
moon, 20–22, 25
Musk, Elon, 95, 96

National Aeronautics and Space Administration (NASA), 6, 8, 12, 21–25, 28, 31, 42, 45, 47, 58–59
NEAR Shoemaker, 58
New Horizons, 47
Newton, Isaac, 18

Oort cloud, 47
Opportunity, 8, 9
OSIRIS-Rex, 59–60

Perlmutter, Saul, 80–81, 88
Philae, 54–58
Planet 9, 48
Pluto, 46–47, 59
Porco, Carolyn, 44

quasars, 78, 80

rockets, 13, 18–20, 22, 25, 26, 92, 95, 96
rogue planets, 66
Rosetta, 54–58
Russia, 13, 25, 97
Ryugu, 60

Sagan, Carl, 42, 97
Saturn, 43, 45, 50, 68
Seager, Sara, 69, 70, 75
Sedna, 47–48
67P/Churyumov-Gerasimenko, 54–58
Soviet Union, 20–21, 25
SpaceX, 92, 95, 96
spectroscopy, 76
Spilker, Linda, 43, 46, 48
Spirit, 8, 9
Sputnik 1, 20
Sputnik 2, 21
Steltzner, Adam, 8, 13

Taylor, Matt, 58
Titan, 45–46
Transiting Exoplanet Survey Satellite (TESS), 69

Ultima Thule, 47
United States, 20–25, 50

Virgin Galactic, 92, 95
Vostok 1, 21
Voyager 1, 40, 42, 45
Voyager 2, 42

World War II, 20
Wright, Ian, 52–54

ABOUT THE AUTHOR

Kathryn Hulick began her career with an adventure. She served two years in the Peace Corps in Kyrgyzstan, teaching English. When she returned to the United States, she began writing books and articles for kids. Technology and science are her favorite topics. She also contributes regularly to *Muse* magazine and the Science News for Students website. She enjoys hiking, painting, reading, and working in her garden. She lives in Massachusetts with her husband, son, and dog.